Studies on Single and Double Actuator Based DC Attraction Type Levitation Systems:

Optimization Techniques

Rupam Bhaduri
Dayananda Sagar University, India

A volume in the Advances in Mechatronics and Mechanical Engineering (AMME) Book Series

IGI Global
PUBLISHER of TIMELY KNOWLEDGE

Published in the United States of America by
IGI Global
Engineering Science Reference (an imprint of IGI Global)
701 E. Chocolate Avenue
Hershey PA, USA 17033
Tel: 717-533-8845
Fax: 717-533-8661
E-mail: cust@igi-global.com
Web site: http://www.igi-global.com

Library of Congress Cataloging-in-Publication Data

Names: Bhaduri, Rupam, DATE- author.
Title: Studies on single and double actuator based DC attraction type levitation systems : optimization techniques / authored by Rupam Bhaduri.
Description: Hershey, PA : Engineering Science Reference, [2023] | Includes bibliographical references and index. | Summary: "The work presented in this book is based on analysis, design and development of DC attraction type levitation system (DCALS). DC electromagnetic levitation systems are receiving increasing attention recently, due to their practical importance in many engineering systems. Presently DCALS is being used in many industrial applications. Extensive research has been carried out over the years throughout the world to design the latest form of the maglev system. There are still many aspects open for research and development in the field of DCALS which require the use of fundamentally interdisciplinary concepts of electromagnetism, electronics and power electronics, mechanical engineering, measurement and control etc"-- Provided by publisher.
Identifiers: LCCN 2023011144 (print) | LCCN 2023011145 (ebook) | ISBN 9781668473887 (h/c) | ISBN 9781668473894 (s/c) | ISBN 9781668473900 (eISBN)
Subjects: LCSH: Lifting magnets--Automatic control. | Magnetic suspension--Mathematics. | Actuators--Mathematical models. | Lifting and carrying. | Electronic control.
Classification: LCC TJ1366 .B53 2023 (print) | LCC TJ1366 (ebook) | DDC 621.8/62015118--dc23/eng/20230403
LC record available at https://lccn.loc.gov/2023011144
LC ebook record available at https://lccn.loc.gov/2023011145

This book is published in the IGI Global book series Advances in Mechatronics and Mechanical Engineering (AMME) (ISSN: 2328-8205; eISSN: 2328-823X)

British Cataloguing in Publication Data
A Cataloguing in Publication record for this book is available from the British Library.

Advances in Mechatronics and Mechanical Engineering (AMME) Book Series

J. Paulo Davim
University of Aveiro, Portugal

ISSN:2328-8205
EISSN:2328-823X

MISSION

With its aid in the creation of smartphones, cars, medical imaging devices, and manufacturing tools, the mechatronics engineering field is in high demand. Mechatronics aims to combine the principles of mechanical, computer, and electrical engineering together to bridge the gap of communication between the different disciplines.

The **Advances in Mechatronics and Mechanical Engineering (AMME) Book Series** provides innovative research and practical developments in the field of mechatronics and mechanical engineering. This series covers a wide variety of application areas in electrical engineering, mechanical engineering, computer and software engineering; essential for academics, practitioners, researchers, and industry leaders.

COVERAGE

- Micro and nanomechanics
- Autonomous Systems
- Nanomaterials and nanomanufacturing
- Intelligent Navigation
- Sustainable and green manufacturing
- Design and Manufacture
- Bioengineering Materials
- Manufacturing Methodologies
- Computer-Based Manufacturing
- Control Methodologies

IGI Global is currently accepting manuscripts for publication within this series. To submit a proposal for a volume in this series, please contact our Acquisition Editors at Acquisitions@igi-global.com or visit: http://www.igi-global.com/publish/.

The Advances in Mechatronics and Mechanical Engineering (AMME) Book Series (ISSN 2328-8205) is published by IGI Global, 701 E. Chocolate Avenue, Hershey, PA 17033-1240, USA, www.igi-global.com. This series is composed of titles available for purchase individually; each title is edited to be contextually exclusive from any other title within the series. For pricing and ordering information please visit http://www.igi-global.com/book-series/advances-mechatronics-mechanical-engineering/73808. Postmaster: Send all address changes to above address.

Titles in this Series

For a list of additional titles in this series, please visit: http://www.igi-global.com/book-series/

Advances in MEMS and Microfluidic Systems
Rajeev Kumar Singh (Department of Mechanical Engineering, Amity University, Noida, India) Rakesh Kumar Phanden (Department of Mechanical Engineering, Amity University, Noida, India) Basant Singh Sikarwar (Department of Mechanical Engineering, Amity University, Noida, India) and J. Paulo Davim (Department of Mechanical Engineering, University of Aveiro, Portugal)
Engineering Science Reference • © 2023 • 331pp • H/C (ISBN: 9781668469521) • US $240.00

Structural Mathematical Modeling Applications in Technological Machines and Transportation Vehicles
Andrey Eliseev (Irkutsk State Transport University, Russia & Irkutsk National Research Technical University, Russia)
Engineering Science Reference • © 2023 • 288pp • H/C (ISBN: 9781668472378) • US $225.00

Handbook of Research on Emerging Designs and Applications for Microwave and Millimeter Wave Circuits
Jamal Zbitou (LABTIC, ENSA of Tangier, University of Abdelmalek Essaadi, Morocco) Mostafa Hefnawi (Royal Military College of Canada, Canada) Fouad Aytouna (LABTIC, ENSA of Tetouan, University of Abdelmalek Essaadi, Morocco) and Ahmed El Oualkadi (LabTIC, ENSA of Tangier, University of Abdelmalek Essaadi, Morocco)
Engineering Science Reference • © 2023 • 542pp • H/C (ISBN: 9781668459553) • US $325.00

Trends, Paradigms, and Advances in Mechatronics Engineering
Mohamed Arezki Mellal (M'Hamed Bougara University, Algeria)
Engineering Science Reference • © 2023 • 299pp • H/C (ISBN: 9781668458877) • US $225.00

IGI Global
PUBLISHER of TIMELY KNOWLEDGE

701 East Chocolate Avenue, Hershey, PA 17033, USA
Tel: 717-533-8845 x100 • Fax: 717-533-8661
E-Mail: cust@igi-global.com • www.igi-global.com

Table of Contents

Preface

The work presented in this book is based on analysis, design and development of DC attraction type levitation system (DCALS). DC electromagnetic levitation systems are receiving increasing attention recently, due to their practical importance in many engineering systems. Presently DCALS is being used in many industrial applications. Extensive research has been carried out over the years throughout the world to design the latest form of the maglev system. There are still many aspects open for research and development in the field of DCALS which require the use of fundamentally interdisciplinary concepts of electromagnetism, electronics and power electronics, mechanical engineering, measurement and control etc.

The electromagnetic levitation system is an interesting problem from the control engineering point of view. It is having an unstable plant with changing parameters, nonlinearities and un-modeled dynamics. DCALS is highly non-linear in nature due to the magnet force-distance-current characteristics. After making appropriate assumptions, the non-linear equations are linearised about an operating point and a linear model has been derived for the proposed system. After estimating the system parameters the transfer functions of the system at different operating points have been determined. DCALS is inherently unstable in open-loop, due to the inverse attractive force vs. air-gap characteristics that essentially places one pole of maglev system on the right half of the 's' plane. The first objective of this dissertation work is to stabilize the overall closed loop system so that the prototype can levitate at a desired air-gap. At the same time DCALS must satisfy the conflicting stringent requirements of a fast response and very less overshoot with good steady state accuracy. So to achieve both stability and good performance for such an unstable system a precision control action is required.

The recent advances in the technology of electromagnetic levitation are attributable almost entirely due to the development of solid state electronic and power semiconductor devices. Power amplifier is the most essential and important part for the implementation of any electromagnetic levitation based project. DCALS critically deserves fast response and the switch mode power amplifiers used for such applications have important role to meet this demand. Different possible switched mode power amplifiers for both single as well as multi-magnet based DC attraction type levitation system have been discussed and a comparative study has been carried out between these power amplifiers based on their structure, working principle, simulated and experimental results. The effect of different parameters of maglev system on the dynamic responses of amplifiers also been observed utilizing frequency response analysis.

Suspension of a sphere is mostly reported in literature. Only single-axis levitation makes complete suspension of any spherical mass due to its structural symmetry. For the complete levitation of a cylindrical rod needs control of at least three degree of freedom movement (roll, pitch and yaw). So the suspension of a long cylindrical rod without tilting is an interesting and challenging task. The concept for the suspension of a cylindrical rod may be utilized in different industrial applications such as induction heating, manufacturing industry, active magnetic bearing, precision instrumentation, mechatronics etc.

Modeling, design, fabrication and testing of single magnet, single axis DC attraction type levitation system (DCALS) for the suspension of a cylindrical rod has been presented. The proposed two-loop scheme is more attractive from control point of view than the mostly reported single-loop position control. The order of the levitated system (plant) has been reduced by one due to the use of a controlled current source as the excitation of the magnet-coil. The analysis and design of the position controller becomes easier after considering the reduced order plant transfer function.

A comparative study between the performances of different classical controllers for the proposed DCALS at different operating points has been carried out and this study suggests a suitable classical controller that will provide fast response without any overshoots and steady-state error which is generally required for such type of critical systems. The simulated controllers have been successfully implemented, thus validating the modeling and controller design procedure.

During experimentation of single-magnet single-axis levitation, it has been observed that the rod gets tilted one side that exerts more levitating force due to non-uniformity of the distributed field flux. Moreover, for some specific industrial applications it is required to levitate such cylindrical rod with better pitching control where both end may be controlled independently. Obviously, for controlling the other degrees of freedom movement of the cylindrical rod, at least two electromagnets are necessary.

Design, implementation and testing of a DC attraction type levitation system where a ferromagnetic cylindrical rod of around 0.122 kg mass is made to remain suspended at the desired operating gap under two electromagnets that have been controlled independently through two identical controllers and the stable levitation of the rod is achieved through SISO control of each air-gap corner. The prototype has been successfully tested and stable levitation has been demonstrated at the desired operating gap.

The basic parameters for DCALS are inductance and resistance of the magnet-coil, input DC link voltage, mass of the payload (object), air-gap between the pole face of the electromagnet and ferromagnetic guide-way or object etc. All these parameters are supposed to get change in real life situation. Moreover, in actual practice the system parameters may vary from the theoretical values due to linear approximation of the electromagnetic force made in the theoretical design, temperature variation, un-modeled dynamics of the electromagnet and other parametric uncertainties. So it is useful to see the performance of the linear controller at the changing situation. The performances of the linear controller with the change in different parameters of the proposed maglev system have been studied. The parameters of the designed controller also get varied due to many reasons. So a sensitivity study of the controller for a DCALS is also being performed.

In this work the proposed unstable maglev systems have been stabilized by linear controllers that have designed by classical methods. It is observed that the closed loop system is stable for the desired operating air-gap but the controllers have a restricted zone of operation. It is found that these controllers provide satisfactory performance in the neighborhood of operating point and performance degradation occurs as the operating point is changed. But in reality there is an urgent need to enhance the capability of the controller so that the prototype can stably levitate over a large air-gap variation. It is aimed to design an optimized controller that will stabilize unstable DCALS and will provide satisfactory performance for a wide range of operating air-gap. Two soft-computing techniques (i) Genetic Algorithm (GA) and (ii) Particle Swarm Optimization (PSO) have been utilized for the said purpose. A comparison has been made between the performances of classically designed controllers with GA and PSO based controllers at different operating positions. It has been observed that the performance of PSO based controller is better than GA and classically designed controller.

Rupam Bhaduri
Dayananda Sagar University, India

Chapter 1
Introduction

ABSTRACT

Magnetic levitation systems are receiving increasing attention recently, due to their practical importance in many engineering systems. There are different applications of magnetic levitation in industry such as levitation of models in a wind tunnel, vibration isolation of sensitive machinery, levitation of molten metal in induction furnaces, levitation of metal slabs during manufacture, bio-medical instrumentation, mechatronics, etc. The suspension of objects in air with no visible means of support due to magnetic force is called magnetic levitation. The development of magnetic materials and magnet design has combined with the development of solid-state electronic devices and their ever increasing current carrying and switching capacity to produce controlled magnetic levitation systems capable of supporting heavy payloads. Magnetic levitation is produced in many different ways.

GENERAL

Magnetic levitation systems are receiving increasing attention recently, due to their practical importance in many engineering systems. There are different applications of magnetic levitation in industry such as: levitation of models in a wind tunnel, vibration isolation of sensitive machinery, levitation of molten metal in induction furnaces, levitation of metal slabs during manufacture, bio-medical instrumentation, mechatronics, etc. (Sinha,1987; Jayawant,1988; Rogg,1984; Yamamura,1976). But the most important applications are in the field of active magnetic bearing (AMB) and transportation systems.

DOI: 10.4018/978-1-6684-7388-7.ch001

Recently a new transport system has been the object of research in many countries. Its main characteristics lie in the fact that it is magnetically levitated, and has guide and propulsion forces generated magnetically also (Sinha,1987; Jayawant,1988; Bittar & Sales,1998). This maglev system has some advantages when compared to the traditional railway system, like absence of contact eliminates frictional loss and aerodynamic noise, reduced component wear, vibration and maintenance cost.

The suspension of objects in air with no visible means of support due to magnetic force is called magnetic levitation. The development of magnetic materials and magnet design has combined with the development of solid state electronic devices and their ever increasing current carrying and switching capacity to produce controlled magnetic levitation systems capable of supporting heavy payloads. Magnetic levitation is produced in many different ways. The techniques currently used are based on one or more of the following principles:

(i) Levitation due to the induced eddy current when there is relative motion between a magnet and a conducting surface
(ii) Levitation due to repulsion between permanent magnets
(iii) Levitation due to force of attraction between ferromagnetic bodies using tuned RLC circuits and alternation currents
(iv) Levitation due to force of attraction between controlled DC electromagnets and ferromagnetic object

But majority of magnetic levitation systems are based on two different principles: attraction (or electromagnetic attraction) and repulsion (or electrodynamic repulsion).

The electrodynamic levitation system (EDS), actuates through repulsive forces generally uses high speed super-conducting magnets (kept in cryogenic container) mounted on the bottom of the moving vehicle and produces the repulsive force due to eddy currents produced in the aluminum guide ways (Sinha,1987; Moon,1994). The eddy current and hence the repulsive force, thus produced, causes levitation beyond a certain threshold speed. The electrodynamic levitation system is inherently stable, but at high speed it possess stability problem due to negative damping (Moon,1994). So some kind of passive damper is required in elctrodynamically levitated vehicle to maintain stability at high speed.

In electromagnetic levitation system (EMS), the levitation is produced due to the attractive force developed between electromagnets and ferromagnetic objects (Jayawant,1988; Rogg,1984). It is inherently unstable and strongly non-linear system. EMS uses the high-power solid-state controls to regulate the current in an electromagnet, and achieves stability through active feedback.

In EMS the electromagnets may be driven either by AC or DC source. Although several experimental systems using AC sources (Moriyama, 2000; Chen & Tsukamoto,1991) have been built, these methods are considered to be suited for applications where mass of the suspended object is small. The severe constraints imposed by eddy-current losses in the magnet and the rather complex control circuitry for power modulation makes the AC method of stabilization inappropriate for heavy payloads (Sinha, 1987). In contrast, the explicit DC method, technically known as the DC attraction type levitation system (DCALS) (Jayawant, 1988), has a considerably simpler configuration with favourable power requirement. In DCALS, the current as well as the attraction force of the electromagnet can be effectively controlled by utilizing a switched mode power amplifier.

The work presented in this book is related to the design and development of DC attraction type levitation system. If a magnet is used to 'float' a ferromagnetic object in a stable position, it is necessary for the gravitational force to be balanced precisely by the attractive force of the magnet. Since the force between the magnet and the floating object, varies as the inverse of the square of the distance between them, a natural conclusion is that a feedback control system is required, so that the strength of the magnet can be adjusted according to the proximity of the object being suspended. Obviously, the prime requirement for the attraction type suspension system is a controlled electromagnet and a contact-less sensor to measure the position of the object as shown in Figure 1 It is clear (Figure 1) that a DC attraction type levitation system (DCALS) primarily consists of four major components: such as (i) electromagnets (ii) power amplifiers, (iii) controllers and (iv) sensors. For successful development of a DCALS one must have an in-depth idea about the individual components. Extensive work is going on over the years throughout the world to design the latest form of magnetic levitation system. There are still many aspects open for research and development in the field of DCALS which require the use of fundamentally interdisciplinary concepts of electromagnetic theory, modeling and simulation, electronics and power electronics, control theory, instrumentation etc.

Figure 1. Simplified closed loop system of DC attraction type levitation system

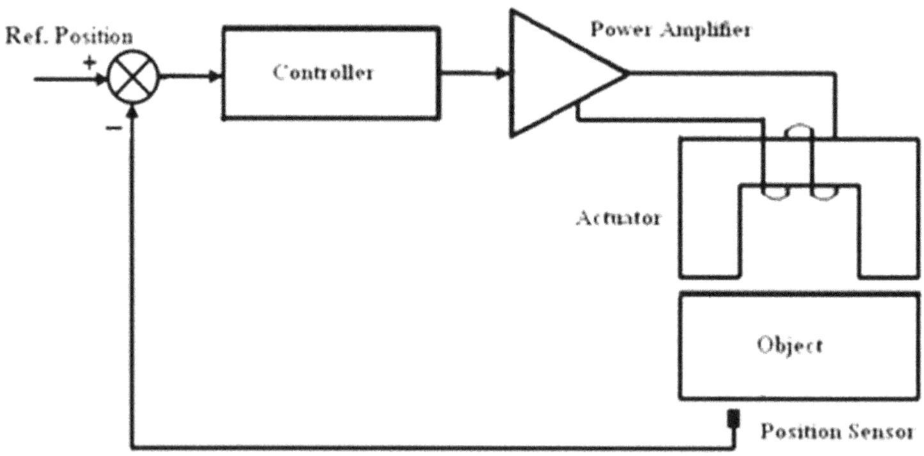

LITERATURE SURVEY

DC attraction type levitation system (DCALS) is a broad term which covers suspension of a small sphere to a large vehicle (maglev train) depending on the type of application. Most researchers have utilized different control algorithms on a simple system such as levitating a small sphere (Shiao, 2001 - Bittar, et.al. 1998). Some researchers have worked on electromagnetic vehicle system (Jayawant, et.al. 1976; Bittar, et.al 1998). In this work, the main emphasis is given to the stabilization of a cylindrical rod for two different systems utilizing different degrees of freedom control.

DCALS is a highly nonlinear system. Most researchers have done the analysis and design of DCALS by taking a linearised model of the system. P.K.Sinha (1987) has given a detailed analysis of the linearised model in his book. In the analysis he has taken controlled voltage source as the excitation to the magnet coil resulting in a third order unstable system. But if one considers a controlled current source as the input to the magnet coil then the order of the system reduces by one. The current controlled levitation system (Banerjee, et.al. 2006 - Banerjee, et.al. 2008) has certain advantages over the voltage controlled system. To determine the transfer function of the model, the basic force constants may be derived from the static characteristics (force vs. distance and force vs. current) of the levitated system (Fig.1.2). Wong (1986) has determined the force constants by taking the inductance profile of the levitated system. The inductance value around the operating point can be approximated by a suitable equation and by using the basic force equation one can find the force constants.

Power amplifiers are necessary for generating appropriate coil-currents in response to the command signal from the controller. Many researchers (Carabelli, et.al. 2000 - Maddaleno, et.al. 1997) have used linear amplifier as a power source for the magnet-coil in DCALS. Recently, there has been some research effort for improvement of the linear power amplifier. Some researchers (Maddaleno, et.al. 1997) have suggested use of class G and modified class G amplifiers in the fields of electromagnetic suspension and active magnetic bearing. Because of high efficiency, switched mode amplifiers are preferred to linear amplifiers in many applications. A single switch based class D chopper can be used for low power applications. The full bridge DC to DC converter circuit consisting of four switches has also been used to excite the magnet-coil. This circuit has better property such as faster dynamics with a smaller current ripple under the same DC link voltage condition. But the full bridge circuit requires more active components and gate drivers and is not the best choice in terms of cost and reliability. Most of the reported switched mode power circuits utilized in electromagnetic levitation use asymmetrical type H-bridge converters with two switches and two diodes (Shirazee, et.al.1995; Banerjee, et.al.2008). The circuit is energy efficient and is ideal for high power applications. For switched mode power supplies, generally pulse width modulation (PWM) technique is used to control the coil current of the magnet (Hurley, et.al. 1997; Boldea, et.al. 1985).

A single switch based switched mode power circuit (Banerjee, et.al. 2008) has been proposed for DCALS. This switched mode power circuit topology can easily be extended to multi-magnet levitation scheme where all the switches of the simplified power circuit may be connected such that their gate drive signals may not need ohmic isolation.

The electromagnetic suspension system is an interesting problem from the control engineering point of view. It is having an unstable plant with changing parameters, nonlinearities and un-modeled dynamics. Different control algorithms (both linear and non-linear) (Barie, et. al. 1996) have been used for magnetic levitation applications. If the design is accomplished by using linear analysis, then classical control based cascade compensation (Sinha, 1977; Hurley, et. al. 1997; Banerjee, et. al. 2006) or state feedback control (Barie, et. al. 1996; Bittar, et. al. 1997) can be implemented. Lead-Lag type compensators have been mostly used (Sinha, 1987; Jayawant, 1988; Sinha, 1979; Banerjee, et. al. 2006) for obtaining the overall closed loop stability of DCALS. The reasons for its popularity can be attributed to simplicity, easy of design and robustness. The other control schemes which have been used in maglev system are: (Bittar, et. al. 1998; Bittar, et. al. 1997), sliding mode (Cho, et. al. 1993), feedback linearizing control (Joo, et. al. 1997; Trumper, et. al. 1997), based control (Lin, et. al. 1998 - Shibukawa, et. al. 1991), gain scheduled control (Kim, et. al. 1994) to name a few. Bittar et. al. 1998, has made a comparison between and classical control applied to an electromagnetically levitated vehicle. Design and analysis of feedback linearizing control for an electromagnetic suspension system has been presented by Joo et. al. 1997. A comparison has been made between sliding mode controller and a classical controller for stabilizing a magnetic levitation system by Cho et. al. 1993. Even though the results are promising, the true advantage of the nonlinear approaches compared to well-designed linear controllers has not yet been established for DCALS having a large payload. An attempt has been made to increase the operating range of the electromagnet by exploiting the concept of piecewise linear control (Banerjee, et. al. 2006) where the nonlinear force-current-airgap relationship of the magnetic suspension system has been successively linearized at different operating points with a suitable controller designed for each of these operating points. The controller design method relies on the concept of approximate model matching (Banerjee, et. al. 2007), and matches certain frequency domain parameters that have a great influence in shaping the transient and steady state time domain (high and low frequency domain) characteristics of the overall closed loop system. An optimization technique utilizing random search method with interval reduction, proposed by Luus and Jaakola has been used for designing the maglev controller that provides stability and performance for an enhanced operating range (Banerjee, et. al. 2008). A robust dynamic sliding mode control (RDSMC) system using a recurrent Elman neural network is proposed to control the position of a levitated object of a magnetic levitation system considering the uncertainties by F. J. Lin et. al. 2009.

To achieve closed loop stability of the system as well as to maintain the desired position of the suspended object it is necessary to apply closed loop feedback control by utilizing some measured information of the object. The controller may exploit such information as position of the object, flux density, current, force, velocity, acceleration or any combination of these. The majority of researchers have used position sensors for the direct measurement of air-gap between the magnet pole-face and object. There are different techniques available in the literature for measuring the air-gap position. Boehm et. al. 1993 has given a comparison of different position sensors including optical sensor, ultrasonic sensor, capacitive sensor, inductive or eddy current sensor,

Hall sensor and laser sensor. They concluded that the eddy current sensor is most useful for magnetic levitation because of its excellent frequency response and very little phase shift.

Position sensing has also been attempted (Yi, et. al. 1995) by indirectly measuring flux density using a Hall-effect sensor. Some researchers have used state feedback control by utilizing different states like position, velocity, and acceleration for feedback purpose. This requires use of an accelerometer and a position sensor. Sensor-less control of magnetic suspension system especially in magnetic bearing applications has also been reported (Vischer, et. al. 1993; Jones, et. al. 1996).

MOTIVATION OF THE WORK

Before highlighting the aim of work it is preferred to present the different problems associated with DC attraction type levitation systems. This may be referred to as the 'problem statement' or 'motivation' of the work. In DC attraction type levitation system (DCALS), the attractive force developed between the electromagnet and ferromagnetic object is directly proportional to the square of the current flowing through the magnet coil and inversely proportional to the square of the air-gap. So DCALS is strongly non-linear system. Apart from these inherent nonlinear characteristics, there are magnetic saturation effects, effect of eddy currents on the moving magnet and hysterisis in the magnetic core.

DCALS is inherently unstable in open-loop, due to the inverse attractive force vs. air-gap characteristics (Figure 2) that essentially places one pole of the transfer function model of the levitated system on the right half of the 's' plane. So it is inevitable to use a closed loop control scheme such that the system will be stable irrespective of any application.

The system parameters of DCALS depend on the air-gap between pole-face of electromagnet and object. With the change of operating air-gap there is a sharp change of coil-inductance and levitating current. These two parameters greatly influence the system transfer-function. It is interesting and challenging task to design a robust optimal control system that will work for a wide operating air-gap.

DCALS critically deserves very fast response and DC-DC switch mode power amplifiers used for such applications have important role to meet this demand. Moreover, power amplifiers will be designed such that it will exhibit the properties like fast dynamics, large bandwidth, ability to carry large payload, energy efficient, high reliability and cost-effective. Depending on applications it is essential to control different degrees of freedom (DOF) movement for a DCALS utilizing multiple actuators, sensors and power amplifiers.

From the discussion it is clear that DCALS is an interesting topic of research and one has to handle the above problems judiciously for the successful implementation of any project based on DCALS.

This research involves work in different fields such as design and fabrication of the experimental set-up including actuator and guide-way, design and fabrication of the power circuit, modeling, design and development of the controller circuit, measurement of signals for feedback purpose e.g., position and current.

Figure 2. Static characteristics of a DC electromagnet with ferromagnetic object

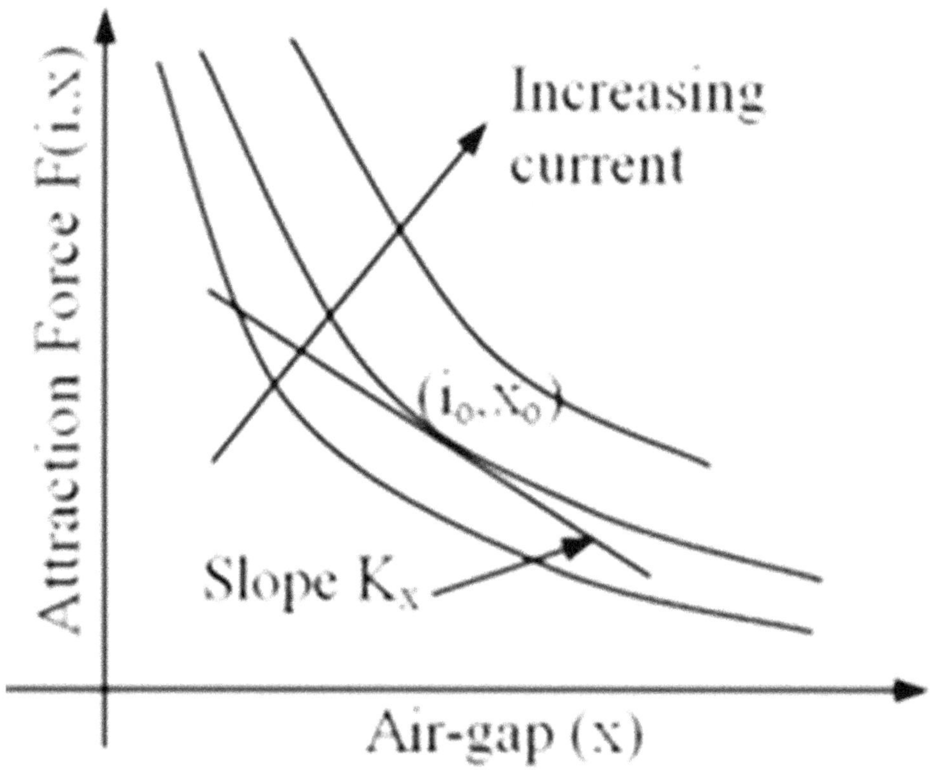

OBJECTIVE OF THE WORK

Only single-axis levitation makes complete suspension of any spherical mass due to its structural symmetry. It is convenient to levitate a ferromagnetic sphere under the force of any electromagnet. Electromagnetic levitation of sphere is mostly cited in literature. For the complete levitation of a cylindrical rod needs at least three degree of freedom movement (roll, pitch and yaw). So the suspension of a long cylindrical rod without tilting is an interesting and challenging task. The concept for the suspension of a cylindrical rod may be utilized in different industrial applications such as induction heating, manufacturing industry, active magnetic bearing, precision instrumentation etc. The different objectives of this work have been summarized as below.

1. To review the work presented in the literature for DC attraction type levitation system.
2. To develop a suitable generalized mathematical model for DC attraction type levitation system.
3. To design, fabrication and testing of a DC attraction type levitation system where a cylindrical rod has to be suspended under an E-core electromagnet with a single degree of freedom in movement.
4. To compare with the different designed controllers for the proposed DCALS at different operating points and to suggest a suitable classical controller that will provide fast response without any overshoots and steady-state error which is generally required for such type of critical systems.
5. To compare between the different topologies of power amplifiers based on their structure, working principle, simulated and experimental results and to suggest a suitable switch mode power amplifier that will have the properties like fast dynamics, large bandwidth, ability to carry large payload, energy efficient, high reliability and cost-effective.
6. To see the effect of different parameters (that are varying during operation) of levitation systems on the dynamic performance of controllers. Sensitivity studies of controller parameters are also important in real-time situation.
7. To design, fabrication and testing of two-magnet based levitation system (for the complete suspension of a ferromagnetic cylindrical rod) that will be controlled through two independent controllers through single input and single output (SISO) control of each air-gap utilizing cascade compensation control scheme with inner current loop and outer position loop.

8. To design optimized controllers utilizing two soft computing techniques (GA and PSO) to make the system (a) operate stably over a wide air-gap range and (b) provide fast stable operation with minimal overshoot and steady-state error irrespective of any operating condition. It is also aimed to compare between the performances of optimized controllers designed by soft computing techniques.

ORGANISATION OF THE BOOK

The book is comprised of eleven different chapters and the organization is given as follows:

Chapter 1

At the outset an introduction of magnetic levitation has been presented. The details of DC attraction type levitation system, literature survey, aim and scope of dissertation and organization of the thesis have been introduced in this introductory chapter.

Chapter 2

The selection of operating point plays a significant role for the design and development of a DC attraction type levitation system. The operating air-gap between the pole-face of the magnet and the object cannot be made arbitrarily too small or too large for many reasons. A detailed description about the selection of operating air-gap has been highlighted in this chapter. It is concluded that the medium zone (air-gap) is best operating area for maglev system. DC attraction type levitation system is strongly nonlinear in nature. Apart from magnetic saturation effects, the nonlinearities which have dominant influence on the stability of the suspended system are magnet force-distance -current characteristics, effect of eddy currents on the moving magnet and hysterisis in the magnetic core. In Chapter 2 a linear mathematical model has been developed for the proposed maglev system. The basic structure of the overall control scheme will depend on the nature of excitation (current source or voltage source) to the magnet-coil. A comparative study of voltage and current controlled levitation scheme has been presented. It is concluded that the current controlled levitation scheme is more advantageous than the voltage controlled scheme.

Chapter 3

Power amplifier is one of the important elements of any DC attraction type levitation system. Due to the advancement in power electronics and control technology it is now possible to develop switching power amplifier which will have the properties like fast dynamics, large bandwidth, ability to carry large payload, energy efficient, high reliability, cost-effective and will be suitable for DCALS.

DCALS critically deserves very fast response and the DC-DC switch mode power amplifiers used for such applications have important role to meet this demand. The coil current for the magnets used in DCALS needs to be precisely controlled to meet the attractive force demand and this calls for a fast DC to DC power amplifier that can be controlled in a closed loop fashion. In the actual work design, fabrication and testing of two different DC attraction type levitation schemes (Chapter 4 & Chapter 6) have been made as a complete project. In this chapter different possible switched mode power circuits for both single as well as multi-actuator based levitation system have been discussed and a comparative study has been made of the different topologies of power amplifiers based on their structure, working principle, simulated and experimental results. A comparison between dynamic performances of different switched mode power amplifiers have been carried out. The effect of different parameters of chopper amplifier (input DC link voltage, duty cycle and switching frequency) and levitation system (resistance and inductance of magnet coil) on the dynamic responses of amplifiers also been observed by frequency response analysis utilizing PSPICE software.

Chapters 4 and 5

Chapters 4 and 5 present design and development of a DC attraction type levitation system where a cylindrical rod is made to remain suspended under one E-core electromagnetic actuator with a single degree of freedom in movement. Detailed description of experimental setup, modeling, design of controllers, hardware circuits and analysis of results have been presented in different sections of this chapter.

The air-gap between the pole-face of E-core electromagnet and the cylindrical rod is measured by a linear inductive type position sensor. Output of the position loop (current reference signal) is compared with the actual coil-current signal sensed by an LEM make (LA-55P) Hall-effect current sensor. A PI controller is used as current controller whose output is used to control the chopper output voltage through PWM control logic. When the electromagnet is excited the rod lifts from platform and the duty ratio of the MOSFET switch varies to control the attractive force developed between the magnet and object. This two-loop scheme is interesting from control point of view than the single loop position control. Controllers have been designed by classical control utilizing root-locus and frequency domain techniques.

One of the objectives of this part of work is to compare with the different designed controllers for the proposed designed DCALS at various operating points and to suggest a suitable classical controller that will provide fast response without any overshoots and steady-state error which is generally required for such type of critical systems. The proposed cascade structure (PI and Lead) enables a better controller design, which would not be the case had a single PID controller been used. This approach decouples the twin requirements of stabilization and performance improvement.

Chapter 6

In this part of work a study has been carried out to see the performance of the proposed control system with the change of different parameters for a DCALS as described in Chapter 4 & 5. Some of the important parameters that are varying during operation of the DCALS are inductance and resistance of the magnet-coil, input DC link voltage, mass of the payload, air-gap between the pole-face of the magnet and ferromagnetic guide-way etc. Moreover, in actual practice the system parameters may vary from the theoretical values due to linear approximation of the electromagnetic force made in the theoretical design, temperature variation, un-modelled dynamics of the electromagnet and other parametric uncertainties. So it is useful to see the performance of the linear controller at the changing environment. On the other hand the sensitivity studies of controller parameters are also important. The study will give an insight for designing any corrective measures with parametric uncertainties for such critical systems.

Chapters 7 and 8

Earlier in Chapters 4 and 5 an attempt has been made for stable suspension of a cylindrical rod with single-magnet single-axis control. It has been observed that the rod gets tilted one side that exerts more levitating force due to non-uniformity of the distributed field flux. Moreover, for some specific industrial applications (like induction heating, manufacturing industry, active magnetic bearing, precision instrumentation etc.) it is required to levitate such cylindrical rod with better pitching control where both end may be controlled independently in decoupled manner. So Chapters 7 and 8 describe design, implementation and testing of a DC attraction type levitation system where a ferromagnetic cylindrical rod of around 0.122 kg mass is made to remain suspended at the desired operating gap under two I-core electromagnetic actuators.

The two actuators are controlled independently through two identical controllers and the stable levitation of the rod is achieved through single input and single output (SISO) control of each air-gap corner. The emphasis of this work is related to the design and development of the switched mode power amplifier cum controller unit for two actuators. The single switch based power circuit simplifies the overall hardware and it can be extended to any number of magnet-coil. A cascade compensation control scheme utilizing inner current loop and outer position loop has been designed and implemented for stabilization of such highly unstable and strongly non-linear system. The prototype has been successfully tested and stable levitation was demonstrated at the desired operating gap.

Chapters 9 and 10

As in Chapters 7 and 8, design, fabrication and testing of two actuator based attraction type levitation system for the suspension of a cylindrical rod has been presented. The controllers have been designed based on classical approach using some thumb rule. It has been observed that the closed loop system is stable for the desired operating air-gap but the controllers have a restricted zone of operation. For a small variation of operating air-gap there is sharp degradation of controller performance. Moreover, the process of controller design is laborious and time consuming.

But it is essential to design an optimized controller that will stabilize unstable DCALS and will provide satisfactory performance for a wide range of operating air-gap. In Chapters 9 and 10 two optimization techniques: (i) Genetic Algorithm (GA) and (ii) Particle Swarm Optimisation (PSO) have been used for the said purpose.

GA has a highly proven track record of optimisation of parameters for different types of control schemes. Here, the work focuses mainly on an optimal control of a proposed two actuator based DCALS scheme, which is composed of a stochastic technique based on Genetic Algorithm (GA).

Deficiencies are there in a GA performance since, the crossover and mutation operations cannot ensure better fitness of offspring because chromosomes in the population have similar structure and their average fitness are high toward the end of the evolutionary process. Whereas PSO has a flexible and well-balanced mechanism to enhance both the global and local exploration abilities. In this part of the work the inertia weighted form of PSO algorithm has been implemented for better convergence. The transient response has been improved by PSO, since optimized parameters of the controllers have occurred more closely to the global optima of the entire search space in the possible solution area. A comparison has been made between the performances of classically designed controllers with GA and PSO based controllers at different operating positions. It has been observed that the performance of PSO based controller is better than GA and classically designed controller.

Chapter 11

In Chapter 11, a conclusion has been drawn summarizing the overall studies. This chapter also suggests scope of future research work on DC attraction type levitation system.

REFERENCES

Banerjee, S., Bhaduri, R., & Prasad, D. (2008). Some aspects of switching power amplifier used in electromagnetic levitation system. *International Journal of Power Electronics*, *1*(1), 33–48. doi:10.1504/IJPELEC.2008.021005

Banerjee, S., Kumar, T. K. S., Pal, J., & Prasad, D. (2008). Controller design for large-gap control of electromagnetically levitated system by using an optimization technique. *IEEE Transactions on Control Systems Technology*, *16*(3), 408–415. doi:10.1109/TCST.2007.906272

Banerjee, S., Prasad, D., & Pal, J. (2006). Large gap control in electromagnetic levitation. *ISA Transactions*, *45*(2), 215–224. doi:10.1016/S0019-0578(07)60191-8 PMID:16649567

Barie, W., & Chiasson, J. (1996). Linear and nonlinear controllers for magnetic levitation. *International Journal of Systems Science*, *27*(11), 1553–1163. doi:10.1080/00207729608929322

Bittar, A., & Sales, R. M. (1997). Control applied to an electromagnetically levitated vehicle. *Proceedings of the IEEE International Conference on Control Applications*, 773–778.

Bittar, A., & Sales, R. M. (1998a). Control for maglev vehicles. *IEEE Control Systems*, *18*(4), 18–25. doi:10.1109/37.710875

Bittar, C. J. J., & Sales, R. M. (1998b). A new approach to the levitation control of an electromagnetic suspension vehicle. *Proceedings of the IEEE International Conference on Control Applications*, 263–267. 10.1109/CCA.1998.728417

Boldea, I., & Nasar, S. A. (1985). *Linear motion electromagnetic systems*. John Wiley & Sons.

Carabelli, S., Maddaleno, F., & Muzzarelli, M. (2000). High-efficiency linear power amplifier for active magnetic bearings. *IEEE Transactions on Industrial Electronics*, *47*(1), 17–24. doi:10.1109/41.824021

Chen, J., & Tsukamoto, O. (1991). Attractive type magnetic levitation system with 3 AC electromagnets. *IEEE International Conference on Industrial Electronics, Control y Instrumentation Proceedings*, *1* 778–783.

Cho, D., Kato, Y., & Spilman, D. (1993). Sliding mode and classical controller for magnetic levitation systems. *IEEE Control Systems*, *13*(1), 42–48. doi:10.1109/37.184792

Hurley, W., & Wolfle, W. (1997). Electromagnetic design of a magnetic suspension system. *IEEE Transactions on Education*, *40*(2), 124–130. doi:10.1109/13.572325

Jayawant, B. V. (1988). Review lecture on electromagnetic suspension and levitation techniques. *Proceedings of the Royal Society of London. Series A, Mathematical and Physical Sciences*, *416*(1851), 245–320. doi:10.1098/rspa.1988.0036

Jayawant, B. V., Sinha, P. K., Wheeler, A. R., Whorlow, R. J., & Willsher, J. (1976). Development of 1-ton magnetically suspended vehicle using controlled d.c. electromagnets. *Proceedings of the Institution of Electrical Engineers*, *123*(9), 941–948. doi:10.1049/piee.1976.0203

Joo, S., & Seo, J. H. (1997). Design and analysis of the nonlinear feedback linearizing control for an electromagnetic suspension system. *IEEE Transactions on Control Systems Technology*, *5*(1), 593–598.

Kim, Y. C. (1994). Gain scheduled control of magnetic suspension system. In *Proceedings of the American Control Conference* (pp. 3127–3131). Academic Press.

Lin, F. J., Chen, S. Y., & Shyu, K. K. (2009). Robust dynamic sliding-mode control using adaptive RENN for magnetic levitation system. *IEEE Transactions on Neural Networks*, *20*(6), 938–951. doi:10.1109/TNN.2009.2014228 PMID:19423437

Lin, J., & Tho, B. (1998). Analysis and -Based controller design for an electromagnetic suspension system. *IEEE Transactions on Education*, *41*(2), 116–129. doi:10.1109/13.669722

Maddaleno, F., Carabelli, S., & Muzzarelli, M. (1997). A modified class G amplifier for active magnetic bearings. *13th Annual Applied Power Electronics Conference Proceedings, 1*, 534–538. 10.1109/APEC.1997.581510

Moriyama, S. (2000). AC magnetic suspension using a differential feedback power amplifier. *Electrical Engineering in Japan*, *131*(2), 61–67. doi:10.1002/(SICI)1520-6416(20000430)131:2<61::AID-EEJ7>3.0.CO;2-Q

Namerikawa, T., & Fujita, M. (2001). Uncertainty structure and μ-synthesis of a magnetic suspension system. *IEEJ Transactions on Electronics. Information Systems*, *121-C*(6), 1080–1087. doi:10.1541/ieejeiss1987.121.6_1080

Oliveira, V. A., Costa, E. F., & Vargas, J. B. (1999). Digital implementation of a magnetic suspension system for laboratory experiments. *IEEE Transactions on Education*, *42*(4), 315–322. doi:10.1109/13.804538

Rogg, D. (1984). General survey of the possible applications and development tendencies of magnetic levitation technology. *IEEE Transactions on Magnetics*, *20*(5), 1696–1701. doi:10.1109/TMAG.1984.1063347

Shiao, Y. S. (2001). Design and implementation of a controller for a magnetic levitation system. *Proceedings of the National Science Council*, *11*(2), 88–94.

Shibukawa, K., Tsubakiji, T., & Kimura, H. (1991). Robust stabilization of a magnetic levitation system. *Proceedings of the 30th Conference on Decision and Control*, 2368–2371. 10.1109/CDC.1991.261609

Shirazee, N. A., & Basak, A. (1995). Electro permanent suspension system for acquiring large air gaps to suspend loads. *IEEE Transactions on Magnetics*, *31*(6), 4193–4195. doi:10.1109/20.489923

Sinha, P. K. (1987). *Electromagnetic suspension, dynamics and control. Peter Peregrinus Ltd.*

Trumper, D. L., Olson, S. M., & Subrahmanyan, P. K. (1997). Linearizing control of magnetic suspension systems. *IEEE Transactions on Control Systems Technology*, *5*(4), 427–438. doi:10.1109/87.595924

Tsukamoto, O., Yasuda, K., & Chen, J. Z. (1988). A new magnetic levitation system with AC magnets. *IEEE Transactions on Magnetics*, *24*(2), 1497–1500. doi:10.1109/20.11528

Wong, T. H. (1986). Design of a magnetic levitation control system – An undergraduate project. *IEEE Transactions on Education*, *29*(4), 196–200. doi:10.1109/TE.1986.5570565

Yamamura, S. (1976). Magnetic levitation technology of tracked vehicles: Present status and prospects. *IEEE Transactions on Magnetics*, *12*(6), 874–878. doi:10.1109/TMAG.1976.1059125

Chapter 2
Modeling of DC Attraction Type Levitation System

ABSTRACT

For the successful implementation of any DCALS, the operating air-gap plays a vital role. Selection of operating air-gap is important to develop a model of any DCALS. The operating air-gap between the pole-face of electromagnet and the object cannot be made arbitrarily too small or too large for many reasons. A very small operating air-gap is not good from safety point of view while there should be enough margins so that, in worst case, the electromagnet does not hit the object. It is always advantageous to select the operating air-gap in between lower and higher gap zone (medium gap). Nevertheless, the final selection of the air-gap depends on the particular system, the type of application, and the payload.

INTRODUCTION

For the successful implementation of any DCALS the operating air-gap plays a vital role. Selection of operating air-gap is important to develop a model of any DCALS. The operating air-gap between the pole-face of electromagnet and the object (Figure 1) cannot be made arbitrarily too small or too large for many reasons (Banerjee & Bhaduri, 2009). A very small operating air-gap is not good from safety point of view while there should be enough margins so that, in worst case, the electromagnet does not hit the object.

DOI: 10.4018/978-1-6684-7388-7.ch002

In most DCALS the inductance variation with air-gap (between pole face of magnet and surface of object) shows three clear zones. Figure 2 shows typical inductance profile of a DCALS. It is seen that in the lower gap zone the inductance varies rapidly, almost exponentially with the gap. In the medium gap zone the inductance variation is inversely proportional to the gap. The inductance does not vary much in the higher gap zone and remains almost constant. Due to high L/R ratio, it is more critical to have a faster control of electromagnetic force in low gap. The large variation of inductance causes wide change in levitated system parameters and the robust controller design becomes a difficult job in low gap. In the higher gap zone the system parameters do not vary widely because of very slow change in the coil-inductance. So robust controller design may be easier in the higher air-gap zone. But in this zone the required current for levitation is very high. The typical characteristic of coil-current versus operating air-gap for a levitated system is shown in Figure 3.

So, in terms of energy consumption the selection of high gap is not desirable. The EMI problem (Boldea, 1985) is much more pronounced in the high gap zone due to large current flowing through the magnet. Moreover, the high value of coil-current becomes a major constraint in the design of actuator, power amplifier and controller. In general, the size of actuator and power amplifier depends on the size of payload. But for a small prototype where the load requirement is supposed to be small, the size of actuator and power amplifier is influenced by the selection of the operating gap. In the case of high current one easy solution is the use of heavy gauge copper wire in actuator and more number of switches in parallel in the power amplifier. But that essentially increases the EMI problem in the control circuit. Selection of the position sensor becomes critical for measuring a large air-gap. The inductive type proximity sensor made by "Contrinex AG Industrial Electronics" of Switzerland is capable of measuring higher air-gaps. However, in electromagnetic applications, this type of sensor suffers from some disadvantages. When a switching amplifier is used, the switching noise from the magnetic suspension system can strongly couple with the probe signals. All these effects get enhanced in higher gap due to the large coil-current. For better stabilization of such an inherently unstable system, the controller mostly used in electromagnetic levitation is PD/Lead type. These controllers are more prone to be affected by noise signals.

Keeping all these points in mind, it is always advantageous to select the operating air-gap in between lower and higher gap zone (medium gap). Nevertheless, the final selection of the air-gap depends on the particular system, the type of application and the payload.

MATHEMATICAL MODELLING OF DC ATTRACTION TYPE LEVITATION SYSTEM

In this work two different prototypes have been designed, fabricated and tested in the laboratory. The simplified diagram of a DCALS has been shown in Figure 1. The mathematical equations of any DCALS are complex and strongly nonlinear. It is aimed to determine a simplified linear model for a DCALS from the basic governing equations.

Figure 1. Simplified diagram of DC attraction type levitation system

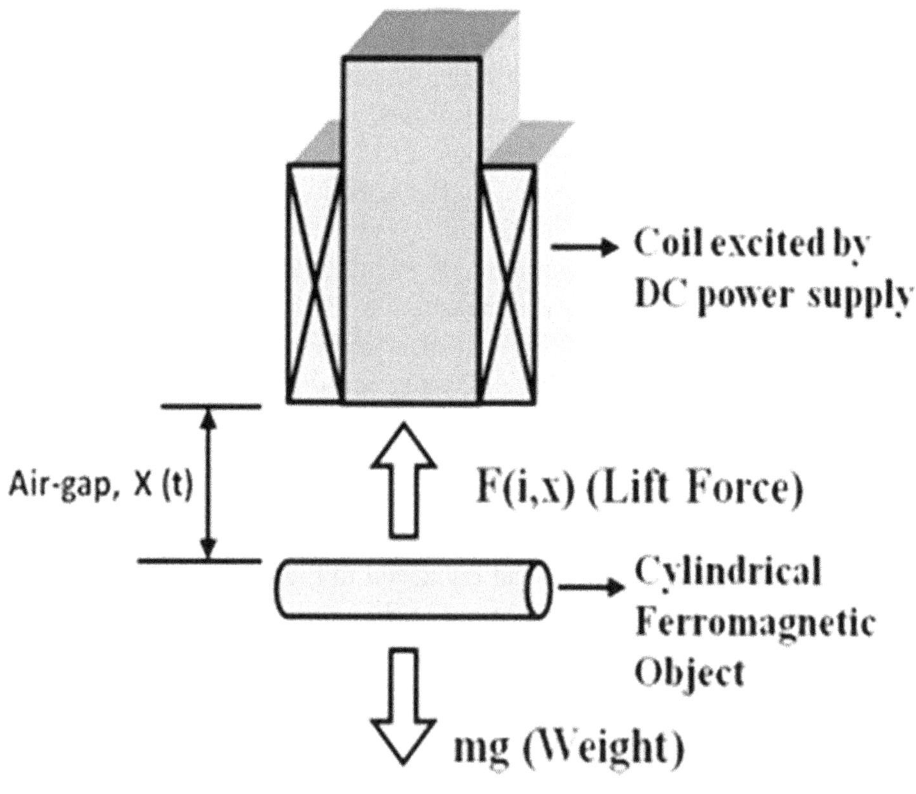

While deriving the governing equations for the mathematical model of the proposed systems, the following simplifying assumptions have been made:

1. The overall magnetic circuit is assumed to be linear. Hysteresis and eddy current effects have been neglected.
2. Leakage flux has been neglected and it is assumed that all magnetic flux generated by the electromagnet passes through the object.
3. The iron permeability of the magnetic material has been assumed to be very high so that the complete magnetizing ampere-turns are spent in the air-gap.
4. Coil inductance is constant about the operating point, and any electromotive force due to motion of the ferromagnetic object can be neglected.

The force of attraction between a ferromagnetic mass and the magnet is non-linear. The force expression is given by

$$F(i,x) = -\frac{d}{dx}\left[\frac{1}{2}L(x)i(t)^2\right] \quad (2.1)$$

Where, $i(t)$ is the current through the coil and $L(x)$ is the inductance of the coil at a particular value of air-gap length (x). The negative sign in the above expression indicates the direction of the generated force. The force is attractive in nature and attempts to decrease the gap length. The suspended steel mass contributes to the inductance of the electromagnet coil. As the body approaches the magnet, the magnet-coil inductance goes up. As the mass moves farther from the magnet, the inductance decreases, reaching a minimal value when the mass is too far. The inductance variation with distance for one system is shown in Figure 2.

The inductance $L(x)$ has its largest value when the object is next to the coil and decreases to a constant value L_C (this is the inductance of electromagnet-coil in the absence of the levitated object) as it is moved to $x = \infty$, for the present purpose (considering operating air-gap in the medium zone) the overall inductance may be approximated as (Wong, 1986; Banerjee & Bhaduri 2009)

$$L(x) = L_C + \frac{L_0 x_0}{x} \quad (2.2)$$

where, L_0 is the inductance at the operating air gap of x_0 (equilibrium position),

Now putting the inductance value from Eqn.2.2 into the force Eqn.2.1 one can write:

$$F(i,x) = C\left[\frac{i(t)}{x(t)}\right]^2 \tag{2.3}$$

where, $C = \dfrac{L_0 x_0}{2}$

Under levitation the force given by Eqn.2.3 is in equilibrium with the gravitational downward pull (mg) acting on the levitating mass.

So, at the equilibrium position (i_0,x_0) the normalized force equation is

$$F(i_0, x_0) = C\left[\frac{i_0}{z_0}\right]^2 = mg \tag{2.4}$$

The dynamics of the object under electromagnet is given by the following equations

$$F(i,x) = mg - m\frac{d^2x(t)}{dt^2} \tag{2.5}$$

From Eqn.2.3 and Eqn.2.5,

$$C\left[\frac{i(t)}{x(t)}\right]^2 = mg - m\frac{d^2x(t)}{dt^2} \tag{2.6}$$

The system dynamic equations are thus nonlinear and hence difficult to analyze. So the equations are linearised about a suitable operating point (i_0,x_0) and the linearised model may be found as described below. If the position of the rod is displaced by an amount $\Delta x(t)$ from the stable point, then let the corresponding change in current and force be respectively $\Delta i(t)$ and $\Delta F(t)$. Here, the coil-current $i(t)$ may be assumed to be composed of two parts: a steady-state component (i_0) which generates the vertical attractive force at an equilibrium point (i_0,x_0), and a much smaller component $\Delta i(t)$which provides the attraction force for balancing any variation around the equilibrium point (i_0,x_0).

From Eqn.2.6, $C\left(\dfrac{i_0+\Delta i(t)}{x_0+\Delta x(t)}\right)^2 = mg - m\dfrac{d^2(x_0+\Delta x(t))}{dt^2}$

$$\Rightarrow C\left(\frac{i_0}{x_0}\right)^2\left(\frac{1+\dfrac{\Delta i(t)}{i_0}}{1+\dfrac{\Delta x(t)}{x_0}}\right)^2 = mg - m\frac{d^2(x_0+\Delta x(t))}{dt^2}$$

$$\Rightarrow C\left(\frac{i_0}{x_0}\right)^2\left(1+\frac{2\Delta i(t)}{i_0}-\frac{2\Delta x(t)}{x_0}-\frac{4\Delta i(t)\Delta x(t)}{i_0 x_0}\right) = mg - m\frac{d^2\Delta x(t)}{dt^2} \tag{2.7}$$

Neglecting higher order terms in Eqn.2.7,

$$\Rightarrow\left(C\left(\frac{i_0}{x_0}\right)^2 + C\left(\frac{i_0}{x_0}\right)^2.\frac{2\Delta i(t)}{i_0} - C\left(\frac{i_0}{x_0}\right)^2.\frac{2\Delta x(t)}{x_0}\right) = mg - m\frac{d^2\Delta x(t)}{dt^2} \tag{2.8}$$

From Eqn.2.4 and Eqn.2.8,

$$\Rightarrow 2C\frac{\Delta i.i_0}{x_0^2} - 2C.\frac{\Delta x.i_0^2}{x_0^3} = -m\frac{d^2\Delta x}{dt^2}$$

$$\Rightarrow 2C\frac{\Delta i(t)i_0}{x_0^2} - 2C.\frac{\Delta x(t).i_0^2}{x_0^3} = -m\frac{d^2\Delta x(t)}{dt^2} \tag{2.9}$$

Taking Laplace transform on both sides of Eqn.9 and after rearranging, the transfer function of the magnetic levitation system is,

$$\frac{\Delta X(s)}{\Delta I(s)} = -\frac{2C\dfrac{i_0}{m.x_0^2}}{\left(s^2 - 2C.\dfrac{.i_0^2}{m.x_0^3}\right)} = -\frac{\dfrac{K_a}{m}}{\left(s^2 - \dfrac{K_X}{m}\right)} \tag{2.10}$$

$$\text{where, } K_a = 2C\frac{i_0}{x_0^2} \tag{2.11}$$

$$\text{and } K_X = 2C.\frac{.i_0^2}{x_0^3} \tag{2.12}$$

are the two force constants which are basically slopes of force vs. current and force vs. air-gap characteristics (Figure 2 in Chapter 1).

Eqn.2.10 represents the linearised plant transfer function of the levitated system, when the magnet-coil is excited by the controlled current source. The transfer function shows that the system is open loop unstable having one pole at $\sqrt{\frac{K_X}{m}}$ on the right half of '*s*' plane. The negative sign in the expression indicates the decrease of object position with the incremental change of coil-current or force.

The instantaneous voltage equation across magnet-coil is given as,

$$v(t) = Ri(t) + L\frac{di(t)}{dt} \tag{2.13}$$

By taking small perturbation model of the Eqn.2.13,

$$v(t) + \Delta v(t) = R\left(i(t) + \Delta i(t)\right) + L\frac{d\left(i(t) + \Delta i(t)\right)}{dt} \tag{2.14}$$

At steady state, Eqn.2.14 becomes,

$$\Rightarrow \Delta v(t) = R.i(t) + L\frac{d\Delta i(t)}{dt} \tag{2.15}$$

Taking Laplace transform of Eqn.2.15,

$$\Rightarrow \Delta V(s) = R.\Delta I(s) + L.s.\Delta I(s)$$

So, the transfer function of the magnet-coil is,

$$\Rightarrow \frac{\Delta I(s)}{\Delta V(s)} = \frac{1}{(R + Ls)} \tag{2.16}$$

When the magnet-coil is excited by a controlled voltage source the transfer function of the magnetic levitation system becomes:

$$G_p(s) = \frac{\Delta X(s)}{\Delta V(s)} = -\frac{\left(\frac{K_a}{m}\right)}{(R + sL)\left(s^2 - \frac{K_X}{m}\right)} \tag{2.17}$$

The linearized state space model consists of three different state variables: object position ($x(t)$), object velocity $\left(\frac{dx(t)}{dt}\right)$ and coil-current $i(t)$. The output variable is taken as the object position ($x(t)$).

From Eqn.2.5, Eqn.2.10 and Eqn.2.17, the state-space representation of the suspended system becomes:

$$\begin{bmatrix} \frac{d\Delta x(t)}{dt} \\ \frac{d^2\Delta x(t)}{dt^2} \\ \frac{d\Delta i(t)}{dt} \end{bmatrix} = \begin{bmatrix} 0 & 1 & 0 \\ \frac{K_x}{m} & 0 & -\frac{K_a}{m} \\ 0 & \frac{K_x}{K_a} & -\frac{R}{L} \end{bmatrix} \begin{bmatrix} \Delta x(t) \\ \frac{d\Delta x(t)}{dt} \\ \Delta i(t) \end{bmatrix} + \begin{bmatrix} 0 \\ 0 \\ \frac{1}{L} \end{bmatrix} \Delta v(t)$$

$$y = \begin{bmatrix} 1 & 0 & 0 \end{bmatrix} \begin{bmatrix} \Delta x(t) \\ \dfrac{d\Delta x(t)}{dt} \\ \Delta i(t) \end{bmatrix} \tag{2.18}$$

The transfer function of the open-loop system then becomes:

$$\frac{\Delta X(s)}{\Delta V(s)} = \begin{bmatrix} 1 & 0 & 0 \end{bmatrix} \frac{Adj(sI - A)}{\det(sI - A)} \begin{bmatrix} 0 \\ 0 \\ \dfrac{1}{L} \end{bmatrix} = -\frac{\left(\dfrac{K_a}{mL}\right)}{s^3 + \dfrac{R}{L}s^2 - \dfrac{K_x}{m}\dfrac{R}{L}} \tag{2.19}$$

Eqn.2.17 shows the linearised plant transfer function of the DCALS when the magnet coil is excited by the controlled voltage source. The above transfer function shows that for a given mass, K_a controls the gain of the open-loop system while K_x controls its poles. The above description either by state space form (Eqn.2.18) or, by transfer function form (Eqn.2.19) is generally sufficient for design and analysis of the suspension dynamics if the reaction surface (guide-way) is stationary and is not affected by the suspension force (Sinha, 1987).

Eqn.2.17 shows the open loop transfer function of the levitated system when the magnet coil is excited by the controlled voltage source. The plant leads to a third order unstable system having a pole on the right half of 's' plane. The denominator term '($sL+R$)' results from the inductance and resistance of the electromagnet. This inertia of the coil introduces a phase lag between a change in object position and a corrective change in electromagnet current. This lag has the effect of making the system unstable. Many researchers (Sinha, 1987; Jayawant, 1988; Hurley & Wolfle, 1997) have worked with this third order unstable plant for getting closed loop stability of the overall electromagnetic suspension system. This necessitates only a single position control loop. But design of position controller is relatively difficult because of higher order transfer function of the plant. Even with a better design of controller, it may be difficult to stabilize the system once it oscillates even at a low frequency, because of the phase lag introduced by the inertia of the coil. Eqn.2.10 suggests that if one uses controlled current source as the input to the magnet coil the order of the plant reduces by one, though the system is still open loop unstable having a pole $\sqrt{\frac{K_X}{m}}$ on the right half of '*s*' plane. To use this plant transfer function for closed loop control of position, essentially one more control loop (current loop) has to be used in conjunction with the slower outer position loop. The inner faster current loop takes care of the inertia of the coil such that dependency of inductance on the displacement is negligible. Design of position controller can be done by only considering the slower outer position loop, which constitutes a simpler second order plant. The inner current loop dynamics may be neglected (as it involves the fast electrical time-constant only) with respect to outer position loop (as it involves mechanical inertia). All these issues will be discussed elaborately in controller design section for proposed DCALS.

DETERMINATION OF MAGLEV TRANSFER FUNCTION

For a given DC attraction type levitation system determination of transfer function of the plant depends on how correctly the two force constants K_a and K_X are measured.

The two force constants are basically the two slopes of the static characteristics (Figure 2 in Chapter 1) of the electromagnet at any operating point. For determining the characteristics of any levitated system, one requires precision arrangement for force measurement with the change in air-gap keeping current in the magnet-coil constant and in the other case; the force measurement is to be done with the change of coil-current for a constant air-gap. Once the characteristics are obtained, the force constants can be found by drawing the tangent at the operating point (i_0,x_0). Therefore, the values of the two constants largely depend on the selection of the operating point. The force constants may also be determined by indirect approach as described below.

In this approach, the inductance profile at different air-gaps between the magnet pole-face and ferromagnetic rod is obtained. The inductance of the levitated system at different air-gap positions can be measured by any standard technique, such as (i) using standard LCR meter at low frequency, around 50 to 100 Hz (ii) measuring impedence of the coil by AC volt-amp method at power frequency (iii) measuring the flux linking the coil by a fluxmeter (iv) by measuring the time constant of the coil-current response to a voltage step. In this work the inductance has found by measuring impedance of the coil by AC volt-amp method at power frequency. The typical inductance profile with air-gap for the proposed levitated system in this work as found experimentally is shown in Figure 2.

As described earlier, in this method of measurement an attempt has been made to approximate the inductance profile preferably about an operating point by a suitable equation. The inductance value around an operating point (in the medium gap range) mostly varies inversely with respect to object position and has been approximated by Eqn.2.2. The two force constants are given by Eqn.2.11 and Eqn.2.12 respectively.

So it is essential to determine the values of i_0 and x_0 to find the values of K_a and K_x. In this case, the desired levitation gap is first fixed. Let us call this the nominal gap x_0. Now using a variable DC supply a slowly increasing voltage is applied to the magnet coil. The coil current is read using an ammeter. When the magnet current is small the upward electromagnetic attraction force is smaller than the downward pull offered due to gravity and hence the object does not move. For this nominal gap the exact current, i_0 at which the object just lifts is noted. Thus we get the DC values of quantities x_0 and i_0 as mentioned in Eqn.2.11 and Eqn.2.12 respectively. The typical characteristic of pick-up current versus air-gap for an electromagnetic levitated system is shown in Figure 3.

Figure 2. Typical inductance profile of DC attraction type levitation system

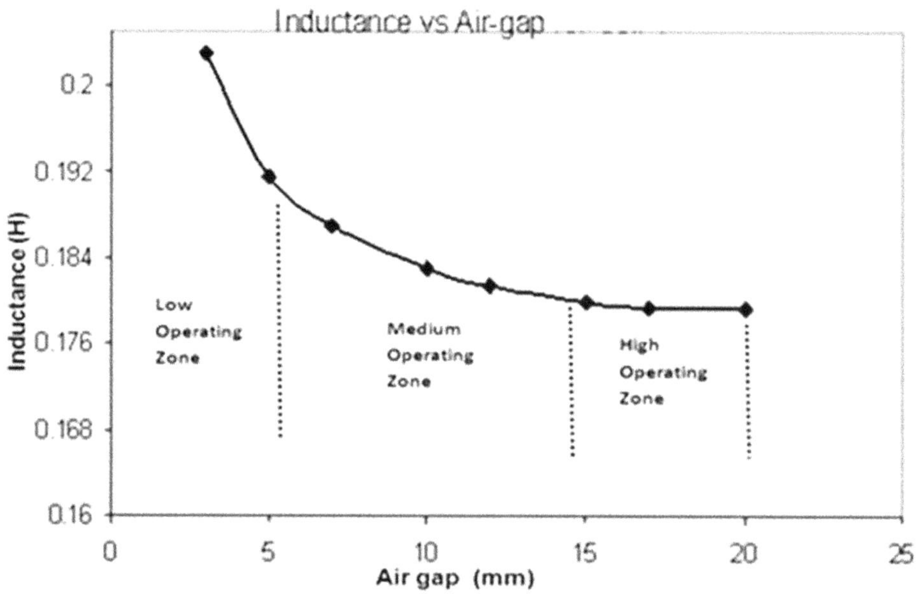

Figure 3. Typical characteristic of pick-up current versus air-gap for a DCALS

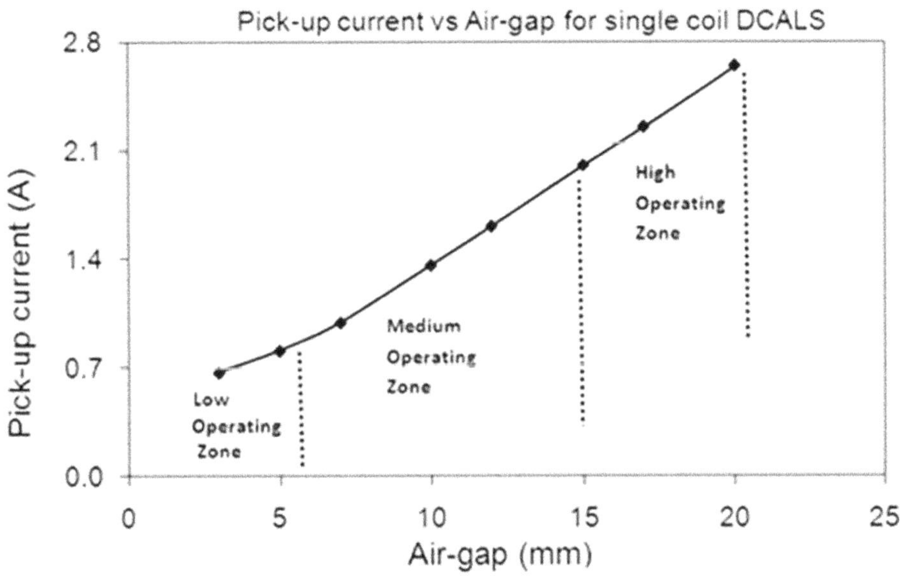

REFERENCES

Banerjee, S., & Bhaduri, R. (2009). An experimental study towards the selection of operating point for a DC attraction type levitation system. *Modeling. Measurement and Control. Part A*, *64*(1–2), 11–18.

Banerjee, S., Sarkar, M. K., Biswas, P. K., Bhaduri, R., & Sarkar, P. (2011). A review note on different components of simple electromagnetic levitation systems. *IETE Technical Review*, *28*(3), 256. Advance online publication. doi:10.4103/0256-4602.81241

Boldea, I., & Nasar, S. A. (1985). *Linear motion electromagnetic systems*. John Wiley & Sons.

Hurley, W., & Wolfle, W. (1997). Electromagnetic design of a magnetic suspension system. *IEEE Transactions on Education*, *40*(2), 124–130. doi:10.1109/13.572325

Sinha, P. K. (1987). *Electromagnetic suspension, dynamics and control*. Peter Peregrinus Ltd.

Wong, T. H. (1986). Design of a magnetic levitation control system – An undergraduate project. *IEEE Transactions on Education*, *29*(4), 196–200. doi:10.1109/TE.1986.5570565

Chapter 3 Some Studies on Switched Mode Power Amplifiers Used in DC Attraction Type Levitation Systems

ABSTRACT

Power amplifier is one of the important elements of any DC attraction type levitation system. Due to the advancement in power electronics and control technology, it is now possible to develop power amplifier which will have the properties of fast dynamics, large bandwidth, ability to carry large payload, energy efficient, high reliability, cost-effective, and suitable for DCALS. Both linear and switching mode power amplifiers have been used in the field of DC attraction type levitation system. In this chapter, different possible switched mode power circuits for both single as well as multi-actuator based levitation system have been discussed, and a comparative study has been made of the different topologies of power amplifiers based on their structure, working principle, simulated, and experimental results.

INTRODUCTION

Power amplifier is one of the important elements of any DC attraction type levitation system. Due to the advancement in power electronics and control technology it is now possible to develop power amplifier which will have

DOI: 10.4018/978-1-6684-7388-7.ch003

the properties like fast dynamics, large bandwidth, ability to carry large payload, energy efficient, high reliability, cost-effective and suitable for DCALS (Moon, 1994; Banerjee, et. al. 2008). Both linear and switching mode power amplifiers have been used in the field of DC attraction type levitation system. In this part of work different possible switched mode power circuits for both single as well as multi-actuator based levitation system have been discussed and a comparative study has been made of the different topologies of power amplifiers based on their structure, working principle, simulated and experimental results.

In the actual work design, fabrication and testing of two different DC attraction type levitation schemes have been made as a complete project. But in this chapter the focus is on switched mode power amplifiers suitable for DC attraction type levitation system.

DCALS critically deserves fast response and the DC-DC switch mode power amplifiers used for such applications have important role to meet this demand. The coil current for the magnets used in DCALS needs to be precisely controlled to meet the attractive force demand and this calls for a fast DC to DC power amplifier that can be controlled in a closed loop fashion. The main objective of this part of work is to compare the dynamic performances between these different power amplifiers so that an insight can be made about the selection of power amplifier where speed of response and bandwidth is the prime criterion. The effect of different parameters of chopper amplifier (input DC link voltage, duty cycle and switching frequency) and levitation system (resistance and inductance of magnet coil) on the dynamic responses of power amplifiers also been observed by frequency response analysis utilizing PSPICE (Banerjee, et. al. 2010). Experimental results for comparing dynamic responses of two different switched mode power amplifiers with a single actuator based maglev system have been produced.

DESCRIPTION OF SWITCHED MODE POWER AMPLIFIERS USED IN DC ATTRACTION TYPE LEVITATION SYSTEMS

The main requirement of power amplifier is to quick control of coil-current as well as attraction force developed between electromagnet and ferromagnetic object. To satisfy this objective, both current rise and fall through the magnet-coil must be fast. This current forcing in both positive and negative directions is essential to achieve the necessary bandwidths of suspension system.

Many power supplies have been reported in the literature for DC attraction type levitation systems. For different maglev systems, both linear and switched mode type solid-state power amplifiers have been proposed which have their own advantages and disadvantages. The linear power supplies produce less switching related noise and the overall electro-magnetic interference is less when compared to the one produced by switched mode power supply (SMPS) circuits. But, in a linear power supply the transistor operates in its active region, incurring a significant amount of power loss. The overall efficiencies of linear power supplies are usually in a range of 30-60% (Mohan, et. al. 1996). For high power magnets, use of linear amplifier will mean un-practically large switch ratings and heat sink ratings. A combination of switched supply and linear amplifier circuit may be thought of to further improve the efficiency of linear amplifier circuits. The switched mode DC to DC power supply is energy efficient (70-90% range). Increased switching speeds, higher voltage and current ratings and relatively lower cost of power devices are the factors that have contributed to the emergence of switching power supplies. A transistor operating in ON/OFF mode has a much larger power handling capability compared to its linear mode. Since a high frequency isolation transformer is used (as compared to a 50-Hz transformer in a linear power supply), the size and weight of switch mode power supplies can be significantly reduced.

A simple class-D type DC to DC chopper circuit of Figure 1 may be the simplest from of power amplifier for the excitation to the magnet coil used in DC electromagnetic levitation system (Sinha, 1987). A full-bridge circuit having four controlled switches (Figure 2) can apply equal amount of positive and negative voltage to the load (magnet coil) while allowing coil current to be bi-directional. A split DC supply shown in Figure 3, for exciting the magnet-coil allows freewheeling and regeneration. Electromagnetic attraction force is, however, independent of the coil-current direction and hence one may as well go for a cheaper asymmetrical-bridge circuit (Figure 4) that allows only one direction of load current (Banerjee, et. al. 2008; Shirazee & Basak, 1995). The single switch based power circuit (Figure 5) may be suitable form of switched mode power amplifier for single and multi-magnet based levitation system.

In this work, a comparison between the dynamic performances for these power amplifiers have been established based on frequency response analysis utilizing PSPICE tool. The effect of different parameters for chopper amplifiers and levitation system on the dynamic responses has also been observed.

Figure 1. Simple class-D chopper circuit

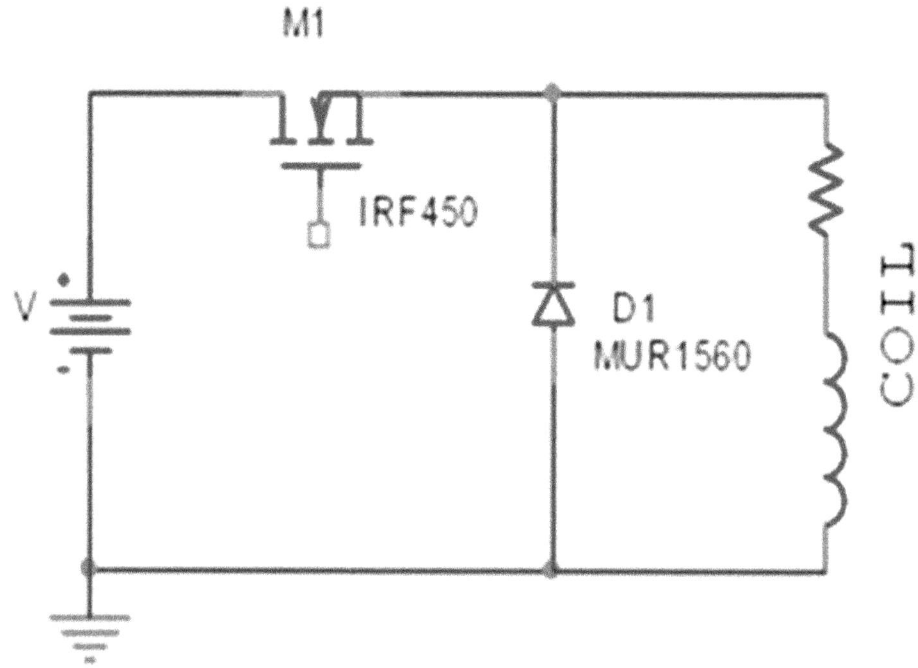

Figure 2. Full-bridge power amplifier

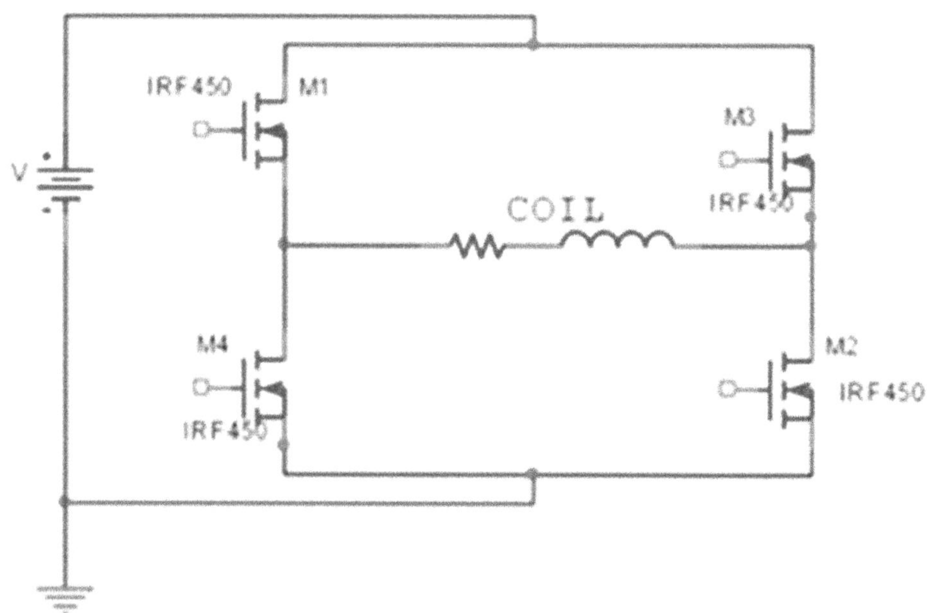

Figure 3. Half-bridge power amplifier

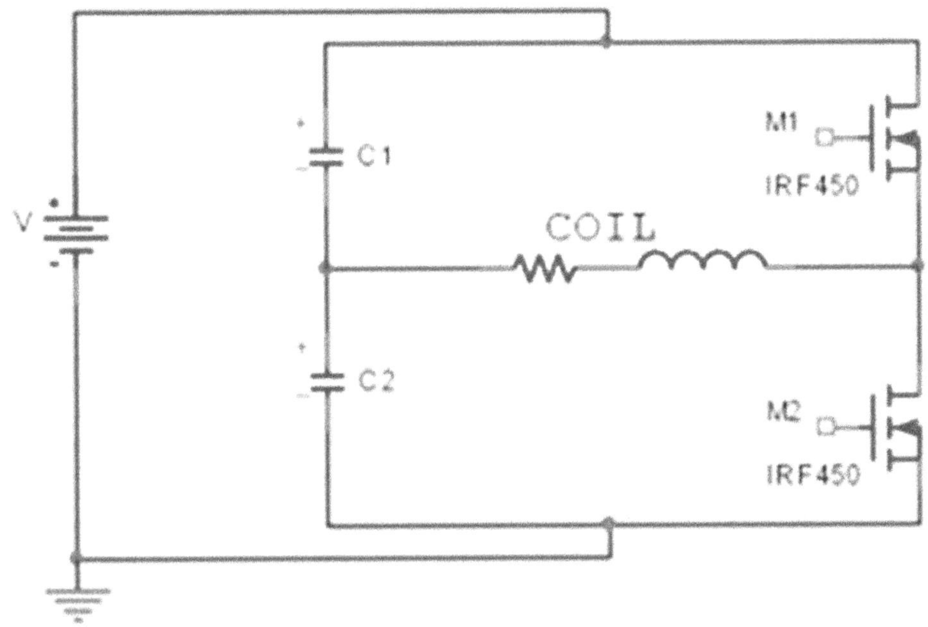

Figure 4. Asymmetrical-bridge converter circuit

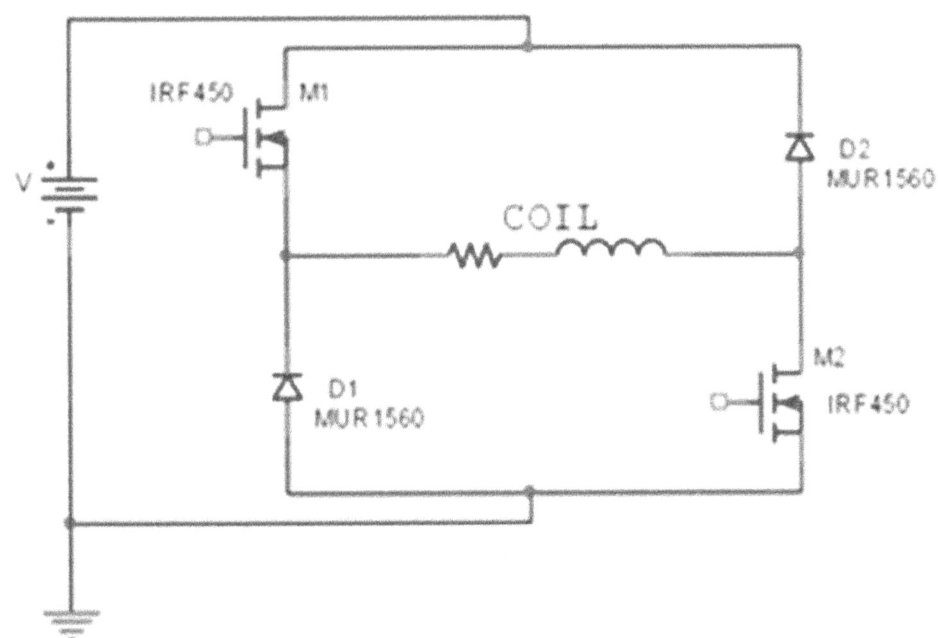

Figure 5. Single switch based chopper amplifier

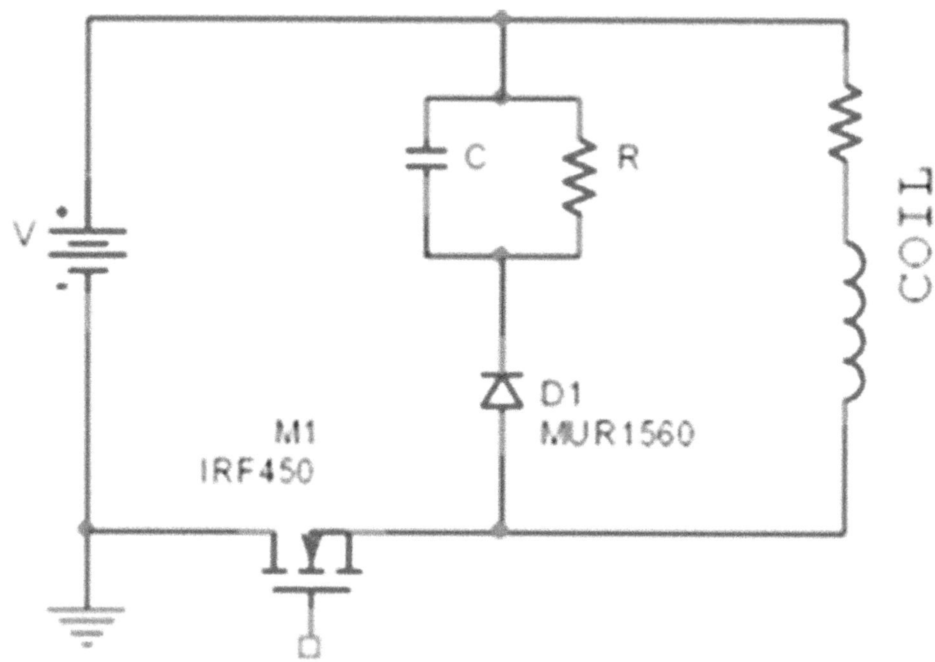

PSPICE SIMULATION RESULTS

The Simulation Program with Integrated Circuit Emphasis (SPICE) circuit simulation tool was first developed in the early 1970s (Cadence Design Systems, 2000). It was written in the FORTRAN programming language and was intended to support the early data entry methods of this period. SPICE was immediately valuable to allow circuit designers to analyze circuit systems, in particular as the complexity of circuits began to expand with the arrival of the first integrated circuits. It is certainly one of the most important tools in Electrical Engineering and is an example of one of the first tools for Computer Aided Design.

SPICE has evolved with many advances in numerical analysis methods for accurate and fast computation and has appeared in many commercial forms. SPICE has been ported to many platforms and the version that operates over the Windows operating system is PSPICE. The first PSPICE users designed circuits with manual circuit drawing tools. Then, inspection of the circuit design was used to generate a text-based description of the circuit design. Even today this still occurs in certain special instances. However, the arrival of graphical drawing tools allows a circuit designer to directly draw circuit schematics using circuit design tools and then special "capture" tools operate to "capture" the schematic and generate the text-based description of the circuit design, automatically. The versions of Pspice Orcad release 15.7 is available to us operate in this way.

PSPICE is the industry-standard way to verify circuit operation at the transistor level before committing to manufacturing an integrated circuit. Board-level circuit designs can often be bread-boarded for testing. Even with a breadboard, some circuit properties may not be accurate compared to the final printed wiring board, such as parasitic resistances and capacitances. These parasitic components can often be estimated more accurately using PSPICE simulation.

This is a great advantage for the engineering design process – this provides the designer with the ability to focus directly on a visual description of their system and obtain circuit operation numerical and graphical results in an automatic, convenient process.

The dynamic response of switching power amplifiers (Figure 1 to Figure 5) have been carried out through frequency response analysis utilizing PSPICE software. Here, the circuit diagram for determining the frequency response of asymmetrical-bridge circuit is presented in Figure 6. For conducting the frequency response one AC voltage source of small magnitude (1 volt) has applied to the load side. Two similar gate pulses have been applied to the two diagonally placed MOSFET switches for symmetrical triggering. As described earlier, when both the switches are ON current flows to the magnet-coil from the source and during OFF period of the switches the stored energy is returned back to the source. The frequency response plot of asymmetrical-bridge converter and half-bridge circuit under similar operating condition are given in Figure 7 and Figure 8 respectively.

Since gain crossover frequency (GCF) determines bandwidth (frequency domain) and the speed of dynamic responses (Ogata, 2000; Clark, 1996), the comparison of different results has been established taking GCF as Y-axis parameter in the different graphs. It has been observed that at any particular operating condition (keeping all parameters same for different converters), the asymmetrical-bridge converter exhibits fastest response while half-bridge circuit is the slowest (Figure 9 to Figure 13). The single switch based circuit (Figure 5) which is easier in construction and suitable for low power application shows slower response than asymmetrical H-bridge converter (Figure 4) under similar operating condition. It is observed that with the increase of duty cycle (Figure 9) and switching frequency (Figure

10) the transient responses of different switching converters become faster. As expected, with the increase of coil inductance the effective time constant of the switching circuit increases making the transient responses slower and vice-versa (Figure 13). The value of coil inductance depends on the operating point of the levitated system.

Figure 6. PSPICE circuit diagram for determining frequency response of asymmetrical-bridge converter

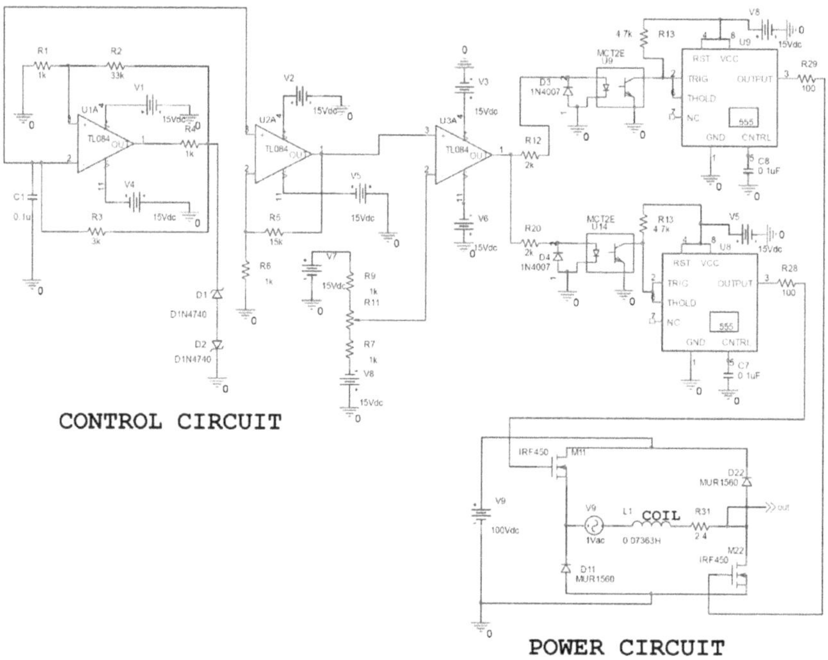

Resistance of the magnet-coil may vary due to the temperature variation after prolonged operation of the experimental set-up. With the increase of coil resistance the response may be faster due to lesser time constant of the magnet-coil of the levitated system. But there is no such appreciable change in GCF (Figure 12) within the specified range of coil-resistance. It is found (Figure 13) that with the increase of input DC link voltage the coil current and hence the m.m.f. can be increased for the same number of turns of the magnet. This method of reducing the overall time constant of the magnet coil is known as voltage forcing (Sinha, 1987). So, with the increase of input DC link voltage, the GCF (dynamic response) of the power amplifiers has increased. The simulated results under different operating conditions for the stated power amplifiers have also been produced in tabulated form (Table 1 to Table 5).

Figure 7. Frequency response plot of asymmetrical-bridge converter at minimum coil-inductance (0.03H) and maximum input voltage (100V), GCF=79.2 kHz

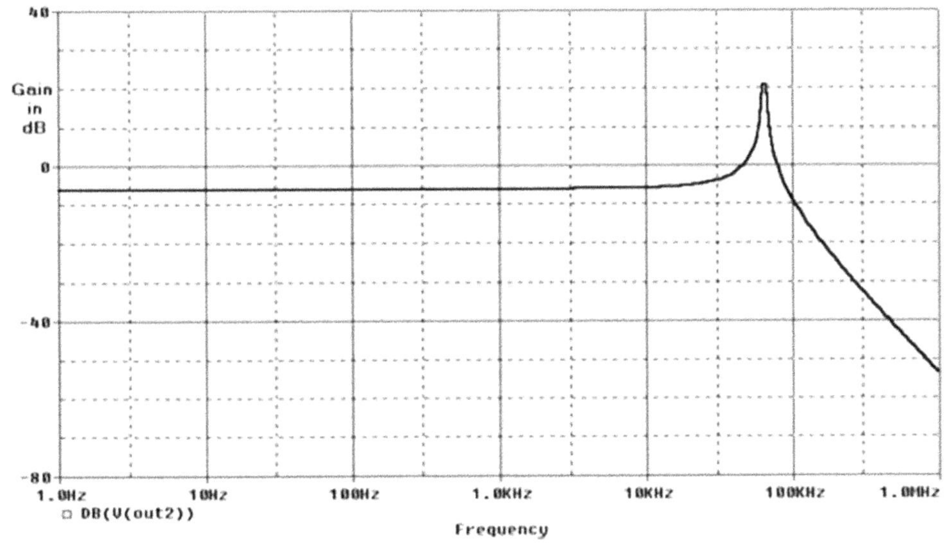

Figure 8. Frequency response plot of half-bridge circuit at minimum coil-inductance (0.03H), and maximum input voltage (100V), GCF=24 KHz

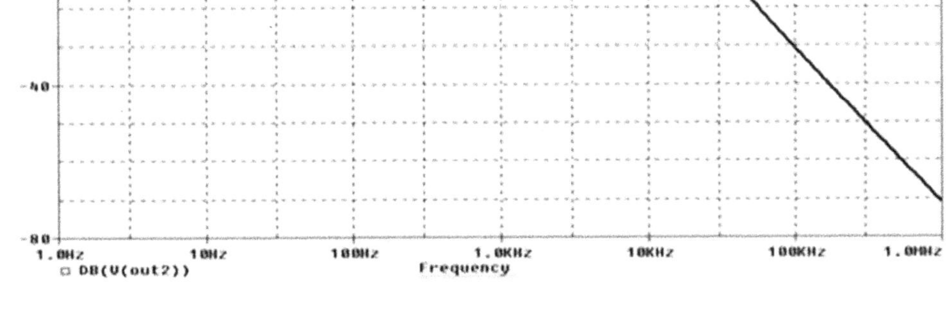

Figure 9. Performance of power amplifiers used in DCALS with duty ratio

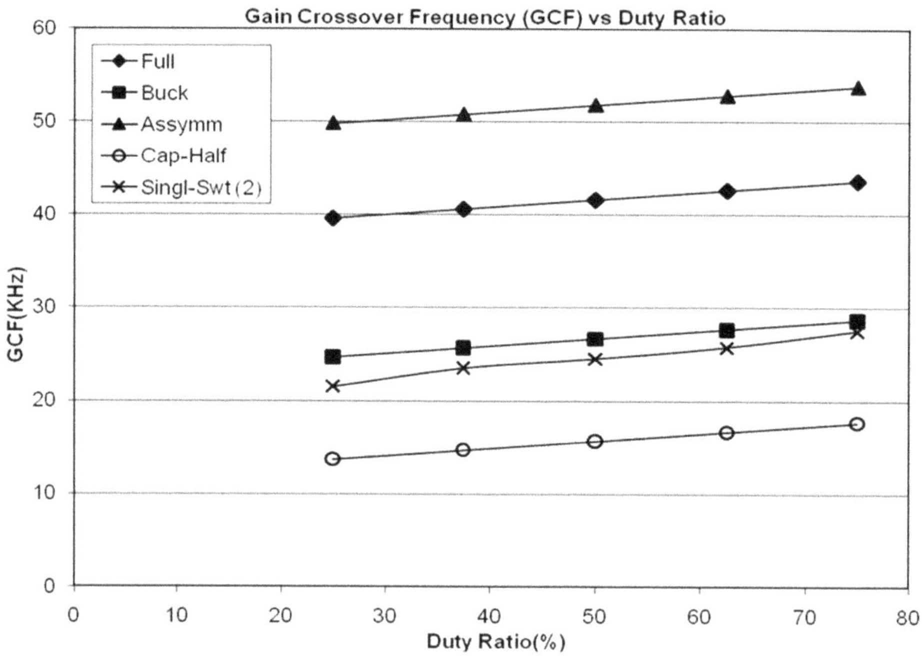

Figure 10. Performance of power amplifiers used in DCALS with switching frequency

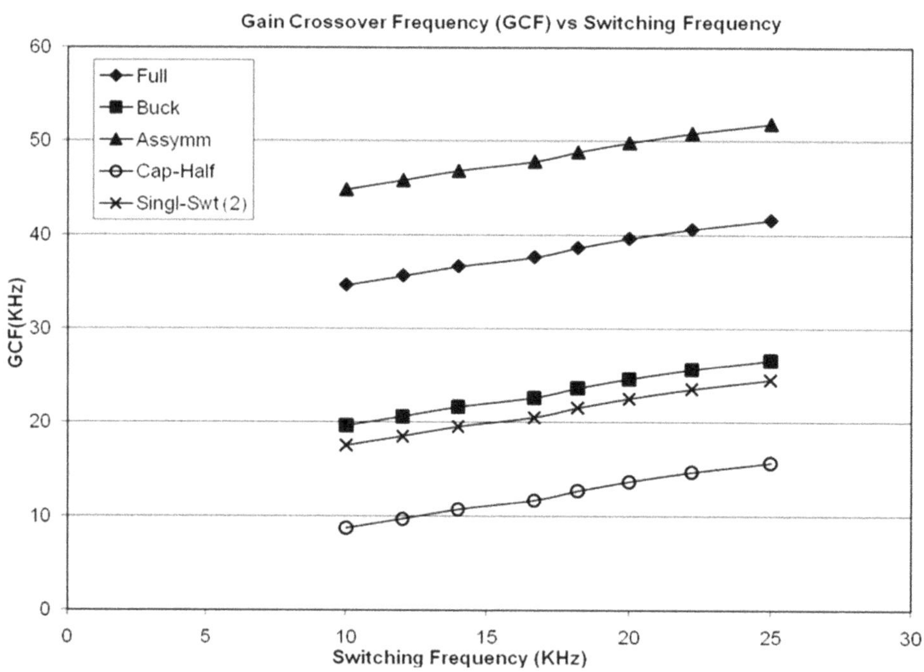

Figure 11. Performance of power amplifiers used in DCALS with coil-inductance

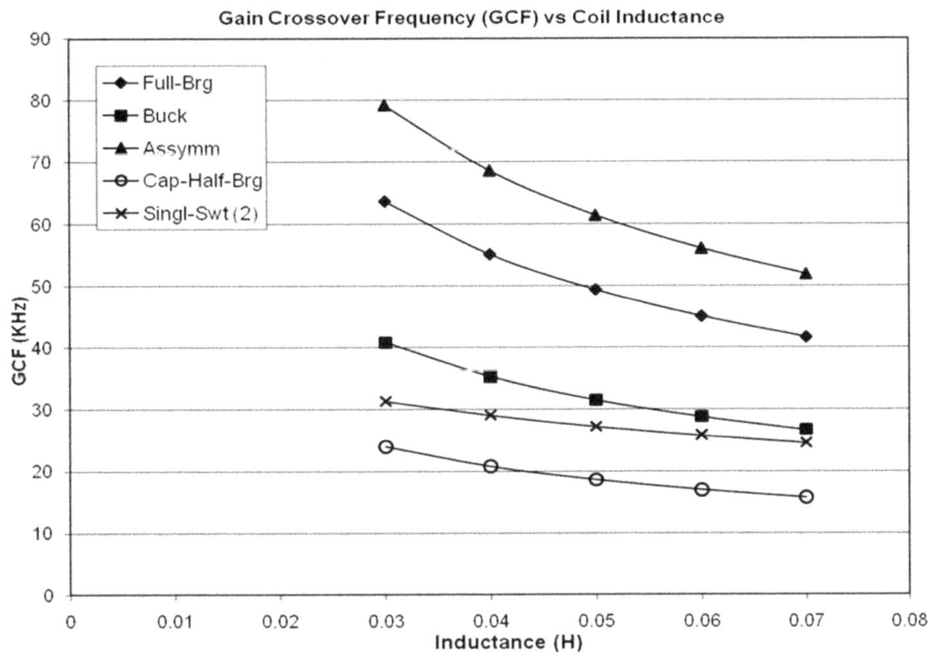

Figure 12. Performance of power amplifiers used in DCALS with coil-resistance

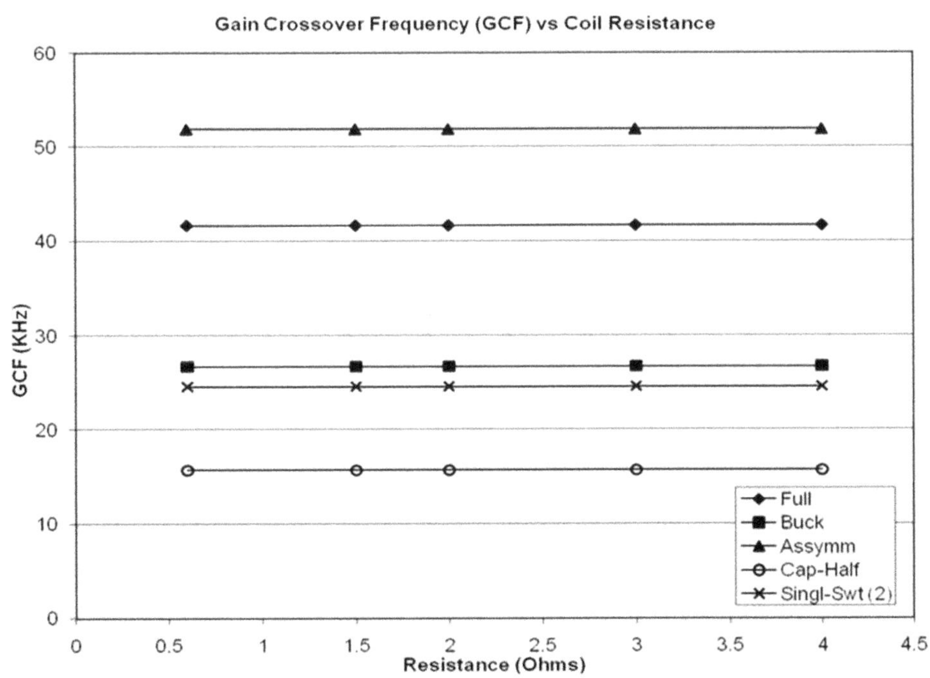

Figure 13. Performance of power amplifiers used in DCALS with input DC link voltage

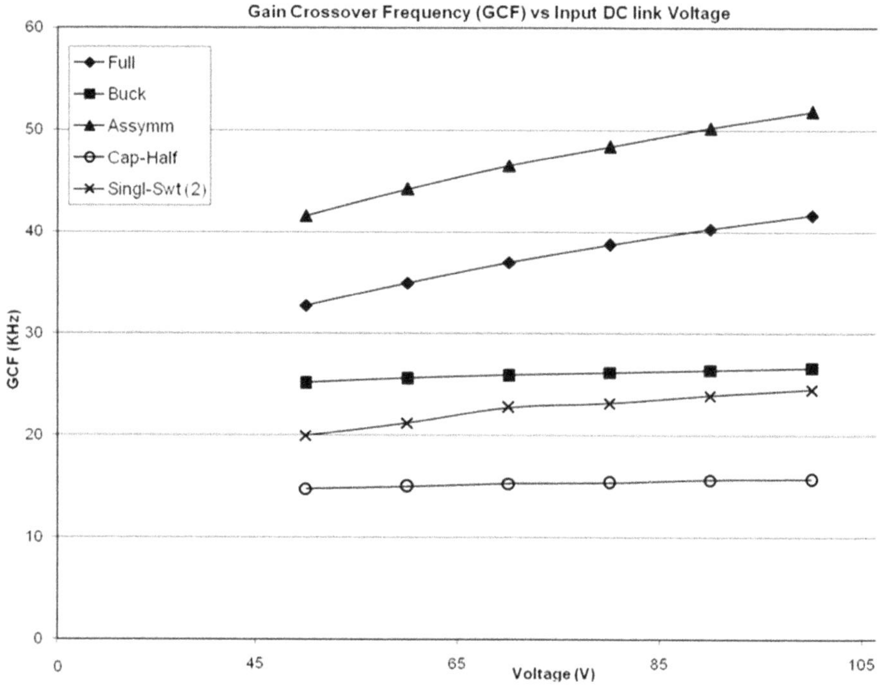

Table 1. Duty ratio vs. GCF for different converters used in DCALS

Full-Bridge	Buck	Asymmetrical-Bridge	Half-Bridge	Single Switch	
GCF (KHz)	GCF (KHz)	GCF (KHz)	GCF (KHz)	GCF (KHz)	Duty Ratio (%)
39.61	24.68	49.85	13.71	21.51	25
40.54	25.08	50.13	14.84	23.46	37.5
41.65	26.68	51.85	15.71	24.55	50
42.65	27.90	52.67	16.61	25.8	62.5
42.98	28.68	53.82	17.79	27.55	75

Table 2. Switching frequency vs. GCF for different converters used in DCALS

Full-Bridge	Buck	Asymmetrical-Bridge	Half-Bridge	Single Switch	
GCF (KHz)	GCF (KHz)	GCF (KHz)	GCF (KHz)	GCF (KHz)	Switching Freq. (KHz)
34.65	19.21	44.81	8.77	17.55	10
35.65	20.60	45.67	9.71	18.15	12
36.65	21.68	45.98	10.71	18.967	14
37.65	22.63	47.45	11.42	20.23	16.67
38.65	22.89	48.58	11.97	21.54	18.2
39.65	23.98	49.34	13.14	21.99	20
40.54	25.08	50.13	14.84	23.46	22.22
41.65	26.68	51.85	15.71	24.55	25

Table 3. Coil inductance vs. GCF for different converters used in DCALS

Full-Bridge	Buck	Asymmetrical-Bridge	Half-Bridge	Single Switch	
GCF (KHz)	GCF (KHz)	GCF (KHz)	GCF (KHz)	GCF (KHz)	Inductance (L) (Henry)
63.668	40.8	79.2	24	31.285	0.03
55.11	35.3	68.58	20.77	29.04	0.04
49.35	31.5	61.35	18.6	27.2	0.05
45.04	28.8	56	16.97	25.77	0.06
41.65	26.68	51.85	15.71	24.55	0.07

Table 4. Coil resistance vs. GCF for different converters used in DCALS

Full-Bridge	Buck	Asymmetrical-Bridge	Half-Bridge	Single Switch	
GCF (KHz)	GCF (KHz)	GCF (KHz)	GCF (KHz)	GCF (KHz)	Resistance (R) (Ohm)
41.65	26.68	51.85	15.71	24.55	0.6
41.65	26.68	51.85	15.71	24.55	1.5
41.65	26.68	51.85	15.71	24.55	2
41.65	26.68	51.85	15.71	24.55	3
41.65	26.68	51.85	15.71	24.55	4

Table 5. Input DC link voltage vs. GCF for different converters used in DCALS

Full-Bridge	Buck	Asymmetrical-Bridge	Half-Bridge	Single Switch	
GCF (KHz)	GCF (KHz)	GCF (KHz)	GCF (KHz)	GCF (KHz)	Voltage (V_{in}) (Volts)
32.75	25.2	41.58	14.77	20	50
35.012	25.67	44.22	15	21.26	60
37	25.94	46.492	15.29	22.8	70
38.76	26.214	48.4	15.4	23.2	80
40.29	26.46	50.21	15.666	23.93	90
41.65	26.68	51.85	15.71	24.55	100

EXPERIMENTAL RESULTS

From the above discussion it is clear that asymmetrical H-bridge converter (Figure 4) is ideal for DCALS applications, while speed of response and energy efficiency is concerned. Though the speed of response of single switch based circuit is slower and the energy efficiency is not comparable to H-bridge circuit, but the circuit is simple having one switch and one diode in contrast to the two switches and two diodes requirement of the asymmetrical H-bridge circuit. There is no need for gate drive isolation as in the case of an asymmetrical-bridge circuit. There will also be less EMI as the number of switches and diodes and hence the number of hard switching per switching cycle will become half. One main disadvantage of the circuit in comparison to the asymmetrical-bridge circuit is that the circuit is not as energy efficient as the full-bridge circuit. Part of the magnet energy is dumped into the capacitor and dissipated into the accompanying parallel resistor. However, in low power applications where simplicity and cost of the circuit may be a factor to be considered vis-à-vis the energy efficiency of the circuit, the proposed circuit may find favour over the asymmetrical-bridge circuit. The advantage of the proposed circuit becomes more apparent in case of a multi-magnet system.

For one single magnet based levitation system both the asymmetrical H-bridge converter and single switch based circuit have been used as power amplifiers and the experimental results have been produced in this section. Figure 14 shows the typical coil voltage and coil current waveform during stable levitation of the object. Coil voltage can be seen to be bipolar with magnitude equal to the input supply voltage. The coil current at this switching frequency is virtually ripple free (ignoring some stray switching noise picked up by the CRO probe).

Figure 14. Coil voltage and coil current during stable levitation of object using asymmetrical H-bridge

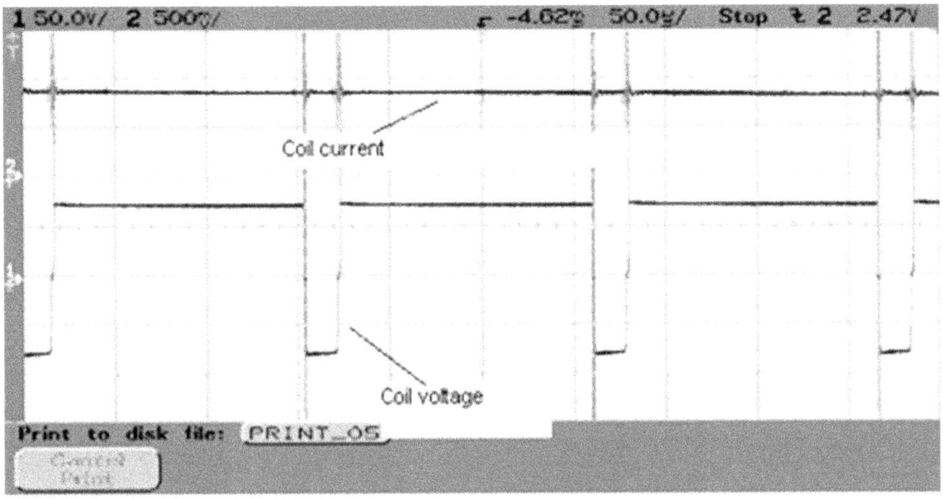

But when the same system is energized by single switch based circuit (Figure 5), the coil-voltage and coil-current waveform is shown in Figure 15. It shows rise and fall of the coil-current along with the coil voltage during stable levitation of the same object. The coil voltage is made equal positive and negative supply voltage with the predefined circuit parameters.

To see the difference between the dynamic responses for the two systems utilizing two power circuits (Figure 4 and Figure 5), during the stable levitation of one object, a small repetitive step voltage of same magnitude and frequency has applied to the position reference input. Figure 16 and Figure 17 shows the position response of the stable levitated system using the asymmetrical-bridge converter and single switch based chopper circuit respectively. It is seen that both the system shows the stable and satisfactory response, but the system with the single-switch based chopper is more under-damped than the other (Banerjee, 2008).

Figure 15. Coil voltage and coil current during stable levitation using single switch circuit

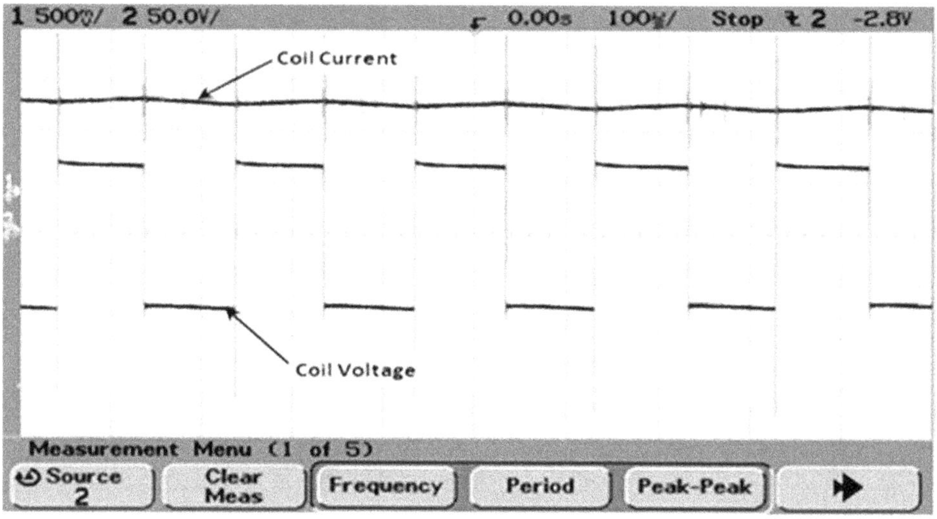

Figure 16. Position response(2V/div) of electromagnet for a square-wave input (1.8v, 1.5Hz, 2V/div) with asymmetrical H-bridge chopper

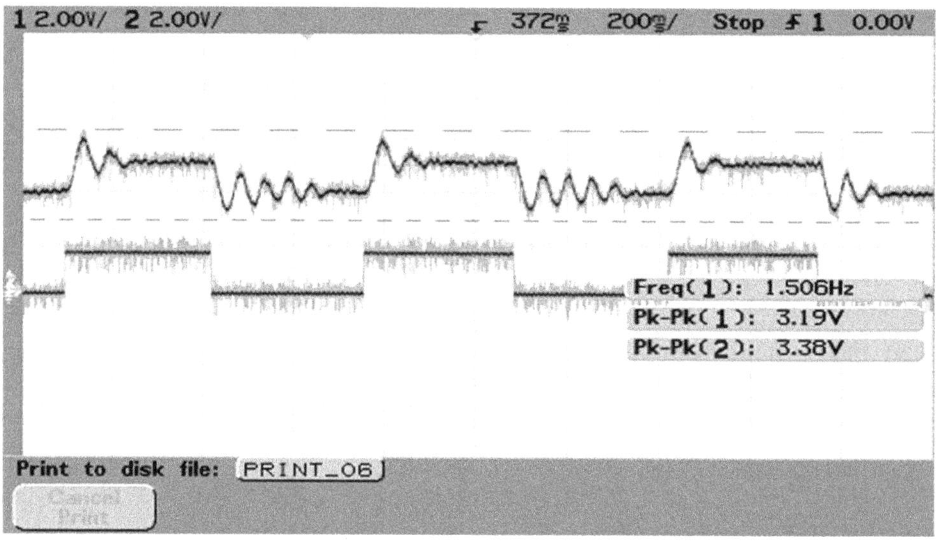

Figure 17. Position response(2V/div) of electromagnet for a square-wave input (1.8v, 1.5Hz, 2V/div) with single-switch based chopper

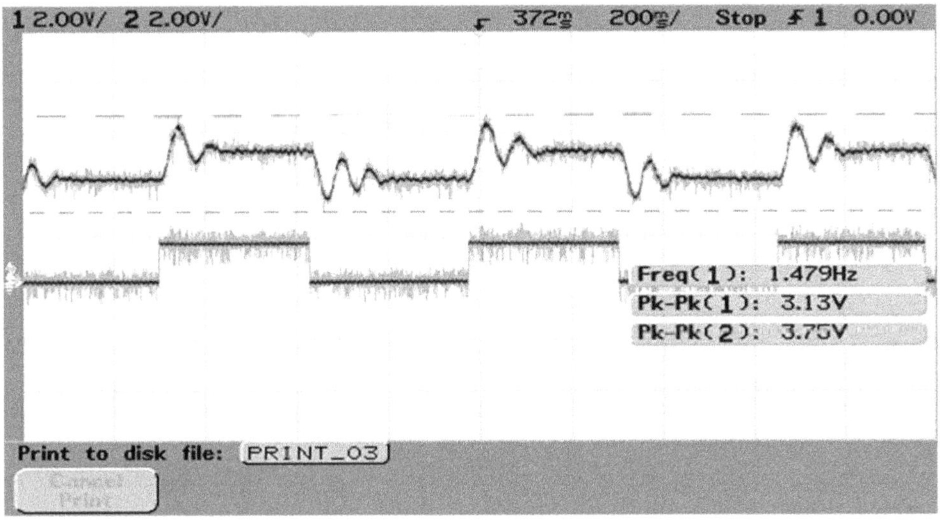

REFERENCES

Banerjee, S., Bhaduri, R., & Biswas, P. (2010, April 19–21). *Frequency response based dynamic performance analysis of switched mode power amplifiers used in electromagnetic levitation systems.* In *IEEE International Conference (PEMD 2010)*, Brighton, UK. 10.1049/cp.2010.0183

Banerjee, S., Bhaduri, R., & Prasad, D. (2008). Some aspects of switching power amplifier used in electromagnetic levitation system. *International Journal of Power Electronics*, *1*(1), 33–48. doi:10.1504/IJPELEC.2008.021005

Cadence Design Systems. (2000). *Orcad capture user's guide* (2nd ed.).

Clark, R. N. (1996). *Control system dynamics*. doi:10.1017/CBO9781139163873

Mohan, N., Undeland, T. M., & Robbins, W. P. (1996). *Power electronic converters: Applications and design* (2nd ed.). IEEE Publications and John Wiley & Sons.

Moon, F. C. (1994). *Superconducting levitation, applications to bearings and magnetic transportation*. John Wiley & Sons.

Ogata, K. (2000). *Modern control engineering* (3rd ed.). Prentice Hall of India.

Shirazee, N. A., & Basak, A. (1995). Electro permanent suspension system for acquiring large air gaps to suspend loads. *IEEE Transactions on Magnetics, 31*(6), 4193–4195. doi:10.1109/20.489923

Sinha, P. K. (1987). *Electromagnetic suspension, dynamics and control*. Peter Peregrinus Ltd.

Chapter 4
Design Fabrication and Testing of a Single Actuator DCALS (Part 1)

ABSTRACT

Suspension of a sphere is mostly reported in literature. Only single-axis levitation makes complete suspension of any spherical mass due to its structural symmetry. It is convenient to levitate a ferromagnetic sphere under the force of any electromagnet. The complete levitation of a cylindrical rod needs control of at least three degrees of freedom movement (roll, pitch, and yaw). So, the suspension of a long cylindrical rod without tilting is an interesting and challenging task. In this book two different kinds of experimental levitation systems have been reported. In this chapter, design and development of a DC attraction type levitation system where a cylindrical rod is made to remain suspended under an E-core electromagnet with a single degree of freedom in movement is explored. Here the E-core electromagnet is fixed, and the levitating cylindrical rod is attracted towards the magnet, and the resulting air-gap is controlled.

INTRODUCTION

Suspension of a sphere is mostly reported in literature. Only single-axis levitation makes complete suspension of any spherical mass due to its structural symmetry. It is convenient to levitate a ferromagnetic sphere under the force

DOI: 10.4018/978-1-6684-7388-7.ch004

of any electromagnet. For the complete levitation of a cylindrical rod needs control of at least three degree of freedom movement (roll, pitch and yaw). So the suspension of a long cylindrical rod without tilting is an interesting and challenging task. The concept for the suspension of a cylindrical rod may be utilized in different industrial applications such as induction heating, manufacturing industry, active magnetic bearing, precision instrumentation, mechatronics etc.

In this book two different kinds of experimental levitation systems have been reported. In this part of work design and development of a DC attraction type levitation system where a cylindrical rod is made to remain suspended under an E-core electromagnet with a single degree of freedom in movement. Here the E-core electromagnet is fixed and the levitating cylindrical rod is attracted towards the magnet and the resulting air-gap is controlled. Detailed description of the experimental setup, modeling, design of controller, hardware circuits and analysis of results are presented in different sections of this chapter.

DESCRIPTION OF THE EXPERIMENTAL SETUP

Electromagnetic levitation requires essentially two necessary subsystems. A primary system used for generating the magnetic field is called actuator and another system is required for shaping or trapping the magnetic flux and is generally termed as guide-way (Moon, 1994). In this prototype an E-core electromagnet is used as an actuator and a ferromagnetic cylindrical rod is used as object. The mechanical structure is such that E-core magnet is fixed with wooden frame and the cylindrical rod is placed below it on a wooden platform (Figure 1). The magnet core is made of laminated sheets of silicon steel and enamelled copper wire is used for winding the coil. The detailed specifications of the electromagnet are given in Table 1. Since the single axis movement of the electromagnet has to be controlled, other degrees of movement have been restricted by providing two tiny rods, fitted vertically between the wooden base and two sides of the rod. The photograph of the complete experimental setup during stable levitation of object is shown in Figure 20 in Chapter 5. When the electromagnet is excited there will be a force of attraction between the magnet pole-face and the ferromagnetic rod. When coil current exceeds the pick-up value, the rod will lift and hit the guide-way. The main aim here is the stable suspension of this rod at some desired operating gap position. To fulfil this objective it is essential to regulate the current of the electromagnet accurately and quickly with the help of position feedback and associated control circuitry. So this levitation process is basically a problem of feedback control system. For sensing the air-gap between the magnet pole-face and the cylindrical rod, an inductive type proximity sensor is used in this work. The sensor with a cylindrical body and a flat sensing surface is fixed with wooden platform where the ferromagnetic rod is positioned. The operating air-gap between the pole face of electromagnet and cylindrical rod may be varied by changing the fitting arrangement of the wooden platform. When the rod lifts the distance between the probe sensing face and the object surface increases resulting in increase in probe output voltage. This arrangement effectively measures the relative distance between the magnet pole-face and the rod surface.

As seen from Figure 1 and Figure 4, output of the position loop (current reference signal) is compared with the actual coil-current signal sensed by an LEM make (LA-55P) Hall-effect current sensor. A PI controller is used as current controller whose output is used to control the chopper output voltage through PWM control logic. When the electromagnet is excited, the rod lifts from platform and the duty ratio of the MOSFET switch varies to control the attractive force developed between the magnet and object. As described earlier, this two-loop scheme is more interesting from control point of view than the single loop position control.

Figure 1. Block diagram for the proposed single actuator based DCALS

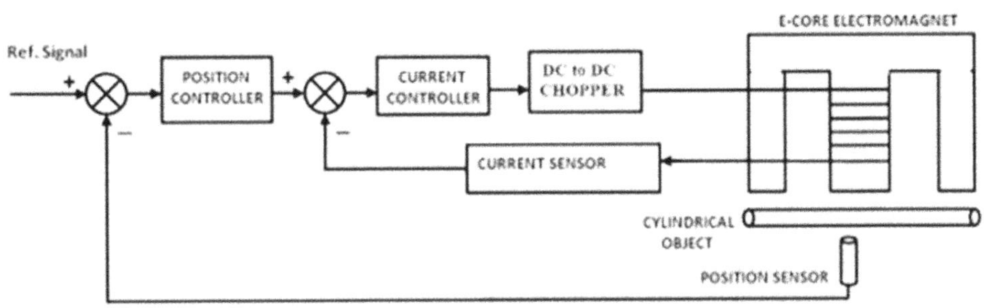

EXPERIMENTAL DETERMINATION OF TRANSFER-FUNCTION FOR PROPOSED MAGLEV SYSTEM

To determine the two force constants (and) as described in Chapter 2, the inductance profile of the levitated system is measured keeping the cylindrical rod under the E-core electromagnet at different gap positions. For this the impedance of the coil is measured by AC volt-amp method at power frequency. The impedance is measured in different voltage levels and then the average value is taken for calculation of the inductance. The same procedure is repeated for different air-gap positions by varying the object position through some mechanical arrangement. The resistance of the coil is known and assumed to be constant throughout the operation. The inductance profile of the levitated system is shown in Figure 2. The ferromagnetic rod contributes to the inductance of the electromagnet coil. As the rod approaches towards magnet-coil inductance goes up and as the rod moves farther from the electromagnet, the inductance decreases, reaching a minimal value when it is too far. This minimum value is, of course, the inductance of the magnet-coil. From the Figure 2, it is obvious that the value of inductance around an operating point (medium range) varies approximately inversely with respect to object position. So the inductance value can be approximated by Eqn.4.1,

$$L(x) = L_c + \frac{L_0 x_0}{x} \tag{4.1}$$

Where is the inductance of the coil in the absence of the ferromagnetic rod and is the additional inductance contributed by the ferromagnetic rod. It may be mentioned here that the above approximation of the inductance is not unique and there may be many different approximations. But incidentally this approximation has worked well in the operating gap region (medium zone) for this particular prototype.

The two force constants are re-written as:

$$K_a = 2C\left(\frac{i_0}{x_0^2}\right) \tag{4.2}$$

$$\text{and } K_X = 2C\left(\frac{i_0^2}{x_0^{\ 3}}\right) \tag{4.3}$$

$$\text{Where, } C = \frac{L_0 x_0}{2} \tag{4.4}$$

After selecting the nominal operating gap the corresponding incremental value of the inductance at this gap is found from Figure 2 and the constant C is calculated by Eqn. 4.4. The corresponding current for gap position can be found from the plot shown in Figure 3. This pick-up current vs. air-gap characteristic of the levitated system is determined as follows. A variable DC supply is used to apply a slowly increasing voltage to the magnet coil. The coil current is read using a PMMC ammeter. For a particular operating gap the exact current at which the ferromagnetic rod just lifts is noted. This pick-up current will be nearly equal to the levitating current at an operating air-gap. This measurement procedure is carried on for different air-gaps between magnet and the object and the results have been shown in the following Figure 3.

Figure 2. Inductance profile of the levitated system

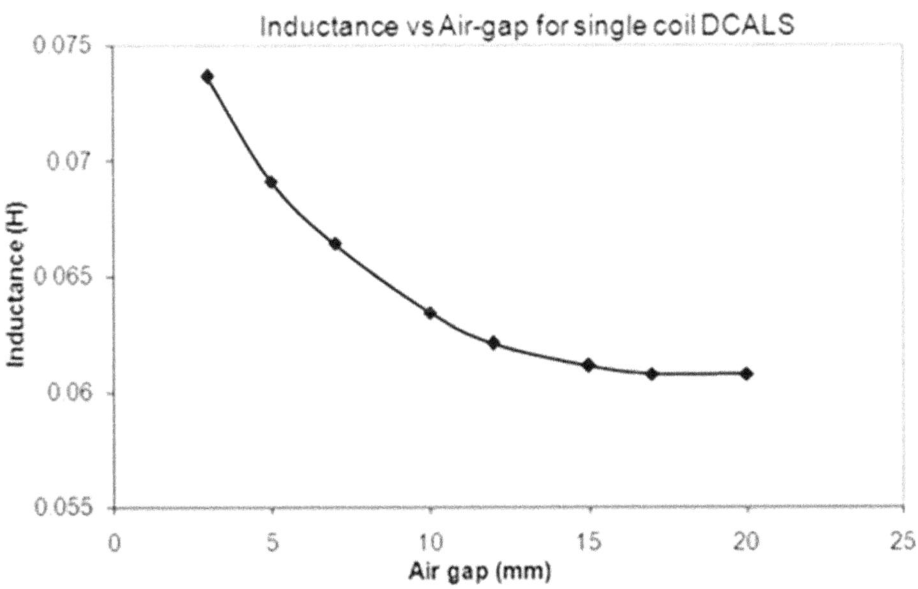

Figure 3. Current profile of the levitated system

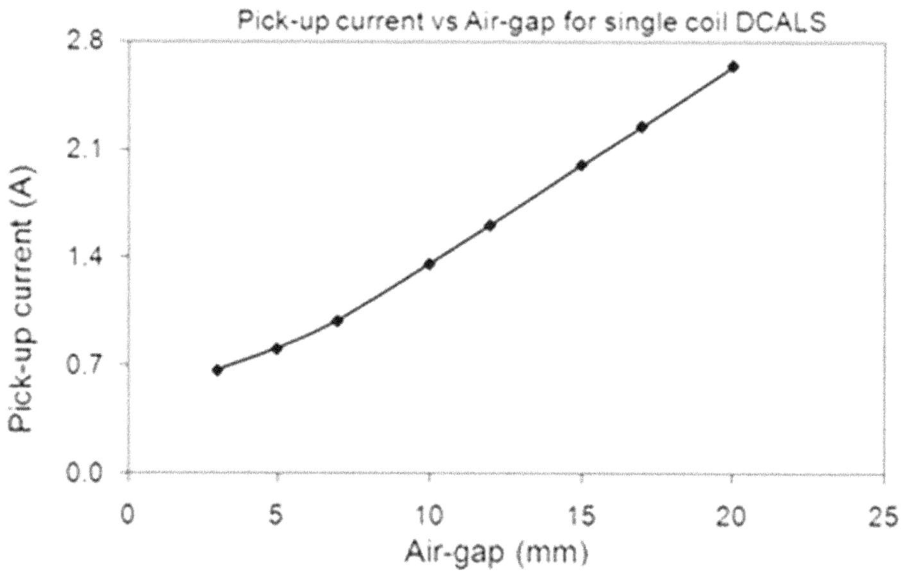

For the present case, controlled current source has been taken as the excitation input to the magnet coil, the magnet kinematics (plant) can be expressed by Eqn.4.5

$$G_p(s) = \frac{\Delta X(s)}{\Delta I(s)} = -\frac{\left(\frac{K_a}{m}\right)}{\left(s^2 - \frac{K_X}{m}\right)} \tag{4.5}$$

After finding the values of the two force constants (K_a and K_X) for a known mass (m), the transfer function of the plant can be determined.

4.3.1 Sample Calculation (for Operating Air-Gap of 10 Mm)

The incremental inductance (L_0) = Total inductance (at air-gap 10 mm) – Coil inductance

= (0.06459H -0.06065H) = 0.00394H.

The force constants are determined as follows:

$$C = \frac{L_0 x_0}{2} = \frac{0.00394 \times 10 \times 10^{-3}}{2} = 0.197 \times 10^{-4}\,; \tag{4.6}$$

$$K_a = 2C\left(\frac{i_0}{x_0^{\ 2}}\right) = \frac{2 \times 0.197 \times 10^{-4} \times 1.36}{10 \times 10 \times 10^{-6}} = 0.5358(N\,/\,A) \tag{4.7}$$

$$K_X = 2C\left(\frac{i_0^2}{x_0^{\ 3}}\right) = K_a\left(\frac{i_0}{x_0}\right) = \frac{0.5358 \times 1.36}{10 \times 10^{-3}} = 72.86(N\,/\,m) \tag{4.8}$$

So the transfer function of the plant (at operating air-gap 10 mm) can be found from Eqn.4.5

$$G_p(s) = \frac{\Delta X(s)}{\Delta I(s)} = \frac{\left(\dfrac{K_a}{m}\right)}{\left(s^2 - \dfrac{K_X}{m}\right)} = \frac{9.237}{\left(s^2 - 1256\right)} = \frac{9.237}{(s + 35.44)(s - 35.44)} \tag{4.9}$$

Table 1. Parameters for single actuator-based DCALS

Parameters	Values
Current sensor gain	1 V/A
Mass of the object	58g
Length of the rod	9cm
Diameter of the rod	10mm
Resistance of the coil	2.4ohm
Inductance of the coil	0.06065H
Position sensor gain	1000 V/m
Input DC link voltage	100V

From Table 2, it is seen that in the lower air-gap region (say at 3 mm to 7 mm) there is a wide variation of the parameters of the plant when compared with respect to the operating air-gap region of (7 mm to 17 mm) due to rapid change of inductance of magnet-coil. Eqn.4.6 to Eqn.4.9 shows that the incremental change of inductance (L_0) of the magnet-coil has a major influence on the system parameters (both the open loop gain and poles) if the current to air-gap ratio is not varied much. Although the inductance variation is not much in the very high air-gap region there is increase in the value of force constants and hence the system parameters with respect to the operating gap due to sharp change in the current/gap ratio. Table 2 shows that with the increase of air-gap (within the operating zone 7 mm-17 mm) the open loop gains as well as poles of the plant decrease in a linear manner. This variation of system parameters with change in air-gap, shown in the Table 2, gives an insight to the realization of controller parameters at different gap positions. It may be mentioned that the variation of system parameters with different air-gaps largely depends on the structure of the system and may vary from system to system.

Table 2. Transfer functions of the single actuator DCALS at different operating positions

Air Gap (mm)	Current (Amp)	Total Inductance (Henry)	Incremental Inductance (Henry)	K_a (N/A)	K_X (N/m)	Transfer-Function of Plant
3	0.58	0.07363	0.01298	2.509	404.07	$\frac{43.26}{(s+83.47)(s-83.47)}$
5	0.84	0.06911	0.00846	1.421	238.72	$\frac{24.53}{(s+64.148)(s-64.148)}$
7	1.03	0.06689	0.00624	0.891	126.06	$\frac{15.368}{(s+46.62)(s-46.62)}$
10	1.36	0.06459	0.00394	0.536	72.86	$\frac{9.237}{(s+35.44)(s-35.44)}$
12	1.73	0.062078	0.001428	0.212	31.41	$\frac{3.652}{(s+23.27)(s-23.27)}$
15	2.02	0.06140	0.00075	0.101	13.60	$\frac{1.74}{(s+15.31)(s-15.31)}$
17	2.40	0.060734	0.000084	0.012	1.694	$\frac{0.2068}{(s+5.402)(s-5.402)}$

DESIGN OF CONTROLLERS

The electromagnetic levitated system is inherently unstable and strongly non-linear in nature. Selection and design of controllers play an important role in the successful implementation of any DCALS. Though a classical controller can be designed based on a linear model, the selection as well as design of such a controller is always a challenging job for ensuring both the stability as well as performance of such critical systems. The basic schematic diagram of the overall closed loop system is shown in Figure 4. It is seen that the system utilizes two-control loops, an inner current loop and an outer position loop similar to a typical servo position system. Here response of the current loop is expected to be faster than the outer position loop, because the inner loop electrical time constant is much less than the outer loop mechanical time constant. The design of control loops starts from the innermost (fastest) loop and proceeds to the slowest loop, which in this case is the outer position loop. The reason to proceed from the inner to the outer loop in the design process is that the gain and time constants of only one controller at a time need to be solved, instead of solving for the gain and time constants of all the controllers simultaneously (Banerjee, et. al. 2006; Krishnan, 2003). The design of the current and position controllers is considered in this section. The controllers are designed based on linear control theory utilizing the root-locus and frequency domain techniques.

Figure 4. Basic block diagram of overall closed loop system

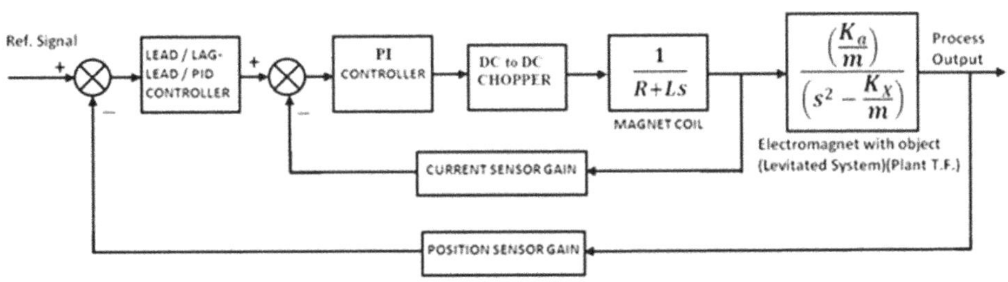

Transfer Functions of the Different Subsystems in the Current Control Loop

To design a suitable current controller for providing better dynamic response it is necessary to model the current control loop accurately. The basic block diagram of the current control loop as shown in Figure 5 consists of different blocks such as current controller, chopper amplifier, magnet transfer-function and current sensor in the feedback path. To design the parameters of the current controller it is necessary to determine the model of the different subsystems of the current control loop. The transfer functions of the different blocks of the current control loop are determined in the following sections.

Current Controller

The current error signal is fed into a controller, which could be a proportional (P), proportional plus integral (PI), or proportional, integral, and differential (PID). The most commonly used controller among them is the PI controller that gives better dynamic performance and ensures zero steady-state error. In this work a proportional plus integral (PI) is used to regulate the current. The transfer function of the current controller is given by

$$G_{cc}(s) = K_p + \frac{K_i}{s} \tag{4.10}$$

Where, K_p and K_i are the proportional and integral gains of the controller.

Chopper Amplifier

The chopper amplifier can be modelled by a gain and a time constant which can be represented by the following transfer function.

$$G_{ch}(s) = \frac{K_{ch}}{(1 + sT_c)} \tag{4.11}$$

Where K_{ch} is the gain and T_c is the time constant of the chopper circuit.

As explained, the chopper duty ratio is decided by comparing PI controller output with a triangular wave with instantaneous magnitude varying between

+5 volts and –5 volts. During each cycle, the chopper switches are on as long as the instantaneous magnitude of PI controller output is more than the instantaneous magnitude of the triangular wave. The time lag in the converter is equal to the average carrier switching-cycle time, i.e., half the period, and is expressed in terms of the PWM switching frequency as $T_c - \frac{1}{2f_c}$; In our case the switching frequency of the converter is taken to be 5 kHz. So the delay angle is about 0.1 ms. so the transfer function of the chopper amplifier becomes:

$$G_{ch}(s) = \frac{20}{(1 + .0001s)} \tag{4.12}$$

After neglecting the very small time constant, the transfer function of the chopper becomes a simple gain of 20.

Electromagnet

From Chapter 2 the instantaneous voltage across the magnet winding can be expressed as

$$v(t) = Ri(t) + \frac{d}{dt}\left[L(x)i(t)\right] \tag{4.13}$$

The inductance is modeled as per the following equation

$$L(z) = L_c + \frac{L_0 x_0}{x} \tag{4.14}$$

One can simplify the analysis by assuming that when the system is properly designed, the suspended object will remain close to its equilibrium position, i.e., $x = x_0$. This leads to the following equation

$$L(x) = L_c + L_0 = L \tag{4.15}$$

L is the total inductance of the coil at a particular gap position and assumed to be constant at the operating point.

So the voltage Eqn.4.13 can be written as

$$V(t) = Ri(t) + L\,\frac{di(t)}{dt} \tag{4.16}$$

Taking the Laplace transform of Eqn.4.16 and after rearranging the transfer function of the magnet-coil appears as:

$$\frac{I(s)}{V(s)} = \frac{1}{(R + sL\ )} \tag{4.17}$$

Experimentally, the transfer function of the coil at the nominal operating point (10 mm) gap is found to be

$$\frac{I(s)}{V(s)} = \frac{1}{(2.4 + 0.06459s)} \tag{4.18}$$

Current Sensor

In this work a LEM make current sensor (LA-55P) was used to measure the actual coil current. The bandwidth of the sensor is very high (1 MHz) and there is perfect (for our purpose) linearity between the actual current and the sensed current signal. A voltage signal proportional to the actual current appears across an output resistor connected between the sensor output and the ground. In this measurement the current sensor was scaled to give an output of 1.0 volts per ampere of coil current. So the feedback path gain (K_c) for the current loop is equal to 1.0.

Design of Current Controller

The block diagram of the current control loop showing the different subsystems in transfer function form is shown in Figure 5.

Figure 5. Block diagram of current loop

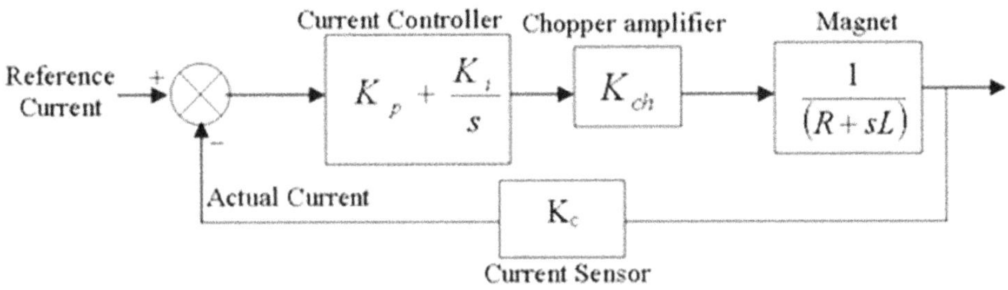

The loop gain is written as

$$G(s)H(s) = K_{ch}K_c(K_p + \frac{K_i}{s}) \times \frac{1}{(R+sL)} = \frac{K_{ch}K_cK_i}{R} \frac{(1+s\frac{K_p}{K_i})}{s} \times \frac{1}{(1+s\frac{L}{R})} \tag{4.19}$$

For the given parameters of the system the gains of the PI controller can be determined both in the frequency and time domain.

Frequency Domain Approach

After converting the above Eqn.4.19 in the frequency domain, from the phase angle criterion one can write:

$$-90^o + \tan^{-1}\left(\frac{\omega K_p}{K_i}\right) - \tan^{-1}\left(\frac{\omega L}{R}\right) + 180^o = \Phi \tag{4.20}$$

Where **Φ**, the phase margin (PM) is assumed to be around 60 deg for good margin of stability. The high value of PM is chosen, because in an experimental implcmcntation additional time lag or saturation decreases the phase margin anyway. By choosing a large PM, stability can be guaranteed for a wide range of operation. The bandwidth of the current feedback loop is to be set higher than the gain crossover frequency of the position control loop

Sample Calculation: (At air-gap 10mm)

At 10 mm air-gap, L = 0.06459 H, R = 2.4Ω

From Eqn.4.19,

$$G(s)H(s) = K_{ch}K_c(K_p + \frac{K_i}{s}) \times \frac{1}{(R + sL)}$$

$$\text{or, } G(s)H(s) = K_{ch}K_c \frac{\left(K_p s + K_i\right)}{s} \times \frac{1}{(R + sL)}$$

$$or, G(j\omega)H(j\omega) = \frac{K_{ch}K_c\sqrt{\omega^2 K_p^2 + K_i^2}}{\omega\sqrt{R^2 + \omega^2 L^2}} \qquad (4.21)$$

It is assumed that phase margin of the current control loop is 60 deg. and the gain crossover frequency (GCF) is around 100 Hz (which is taken at least ten times of the allowable vibration frequency of suspension system that is around 10 Hz)

So, P.M = $180^0 + \Phi$

$$\text{or, } -90^0 + \tan^{-1}\left(\frac{\omega K_p}{K_i}\right) - \tan^{-1}\left(\frac{\omega L}{R}\right) + 180^0 = 60^0$$

$$\text{or, } \frac{K_p}{K_i} = 2.42 \times 10^{-3} \qquad (4.22)$$

Now, $|G(s)H(s)|=1$ where, gain cross over freq. GCF in Hz.=100 Hz.

$$\text{Or,, } \frac{K_{ch}K_c\sqrt{\omega^2 K_p^2 + K_i^2}}{\omega\sqrt{R^2 + \omega^2 L^2}} = 1$$

$$\text{Or, } \frac{20\sqrt{394384K_p^2 + K_i^2}}{628\sqrt{5.76 + 394384.(0.4225)}} = 1 \qquad (4.23)$$

After solving the Eqn.4.22 and Eqn.4.23 the value of K_p(=1.711) and K_i(=707.33) is determined.

From Eqn.4.19 the loop gain becomes,

$$G(s)H(s) = \frac{5894(1 + 0.0024s)}{s(1 + 0.0269s)} \quad (4.24)$$

The Bode plot of the current control loop with designed PI controller is shown in Figure 6. Here the achieved phase margin is 60 degree at a gain crossover frequency of 100 Hz which validates the current controller design.

Figure 6. Frequency response of the current loop

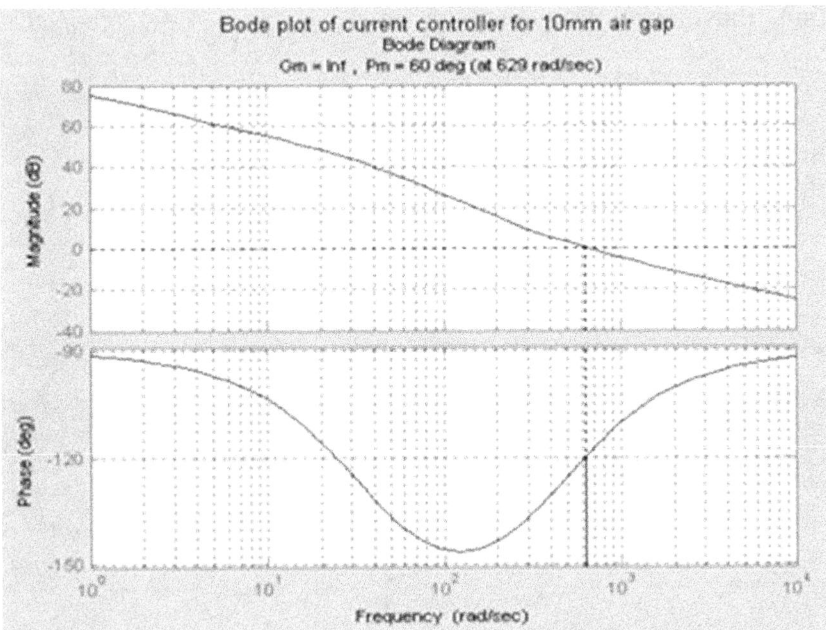

Time Domain Approach

The characteristic equation of the current loop is

$$1 + G(s)H(s) = 1 + \frac{K_{ch}K_c(sK_p + K_i)}{s(R + sL)} = 0 \quad (4.25)$$

or, $s(R+sL) + (sK_p+K_i)K_{ch}K_c = 0$;

$$L\left[s^2 + s\left\{\frac{R}{L} + \frac{K_pK_{ch}K_c}{L}\right\} + \frac{K_iK_{ch}K_c}{L}\right] = 0$$

Since L is not equal to zero,

$$s^2 + s\left\{\frac{R}{L} + \frac{K_pK_{ch}K_c}{L}\right\} + \frac{K_iK_{ch}K_c}{L} = 0 \qquad (4.26)$$

By comparing with the prototype second order time domain characteristic equation, the natural frequency becomes, $\omega_n = \sqrt{\frac{K_iK_{ch}K_c}{L}}$ rad/sec. and damping ratio $\zeta = \frac{1}{2}\frac{\left\{\frac{R}{L} + \frac{K_pK_{ch}K_c}{L}\right\}}{\sqrt{\frac{K_iK_{ch}K_c}{L}}}$

Figure 7. Unity step response of current loop

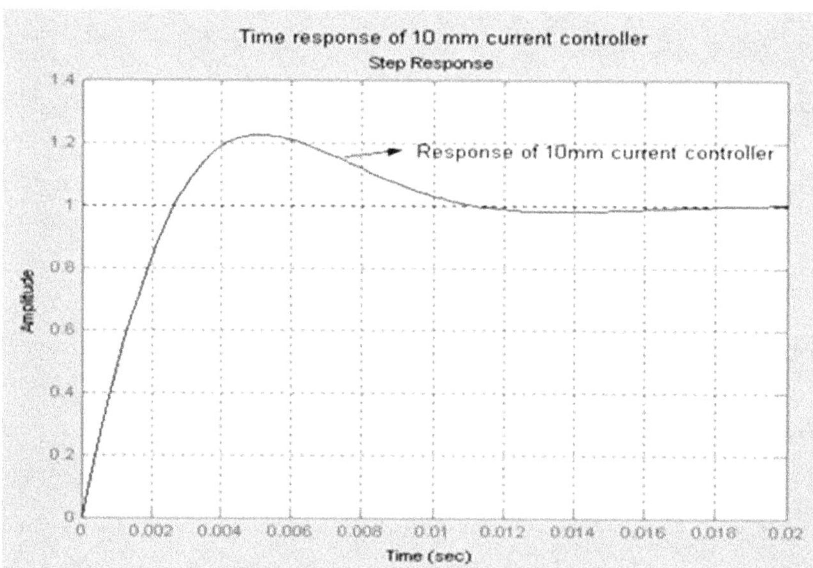

The unit step response of the current control loop is given in Figure 7. As expected the plot shows a good dynamic response having an overshoot of 22% and zero steady state error as expected.

The natural frequency is generally taken as 60% to 80% of the gain crossover frequency of the system. After assuming some standard values of natural frequency and the damping ratio one can determine K_p and K_i; to have the same values of gain of the controller i.e., for K_p=1.711 and K_i=707.33 one should take the values of ω_n= 72 Hz and ζ= 0.606.

The bandwidth of the current loop may also be determined by the following equation

$$\omega_b = \omega_n \left[\sqrt{1 - 2\zeta^2 + \sqrt{(2 - 4\zeta^2 + 4\zeta^4)}} \right] \tag{4.27}$$

The current controllers for different air-gap positions have been determined adopting similar technique. Table 3 shows the performance of current controllers at different air-gaps. It is seen that there is no such appreciable change of gain and time constant of fast acting current controller for a large variation of air-gap (3mm to 17mm).

Transfer Functions of the Different Subsystems in the Position Control Loop

After designing the current controller the current loop is reduced to a single block that simplifies the design of the position control loop. For designing the slower position loop controller often the faster inner loop dynamics is neglected or, approximated to a first order model. In this way one can get an idea about the range of the parameters of the position loop controller. However, to get more accurate values of the outer loop controller parameters for obtaining the desired response, it is always advisable to consider the complete model of the inner loop if the order of the same is not high. The block diagram of simplified position control loop as shown in Figure 8 consists of different blocks such as position controller, current loop, levitated system (plant), and position sensor in the feedback path. To design the position controller it is necessary to obtain the model of these subsystems

Table 3. Performance of current control loop at different operating air-gaps

Air Gap (mm)	Current Controller (PI)	Gain Margin (db)	Phase Margin (Deg)	Maximum Overshoot	Rise Time (sec)	Peak Time (sec)	Settling Time (sec)	Steady State Error
3mm	$1.954\frac{(s+407.6)}{s}$	Infinite	60	22.7%	0.00204	0.00512	0.00143	0
5mm	$1.816\frac{(s+410.5)}{s}$	Infinite	60	22.6%	0.00204	0.00512	0.0143	0
7mm	$1.763\frac{(s+412.4)}{s}$	Infinite	60	22.5%	0.00204	0.00511	0.0143	0
10mm	$1.711\frac{(s+413.8)}{s}$	Infinite	60	22.4%	0.00204	0.00510	0.0142	0
12mm	$1.628\frac{(s+416.5)}{s}$	Infinite	60	22.4%	0.00205	0.00509	0.0142	0
15mm	$1.607\frac{(s+417.3)}{s}$	Infinite	60	22.3%	0.00204	0.00507	0.0141	0
17mm	$1.573\frac{(s+418.4)}{s}$	Infinite	60	22.3%	0.00205	0.00509	0.0142	0

Figure 8. Block diagram of simplified position control loop

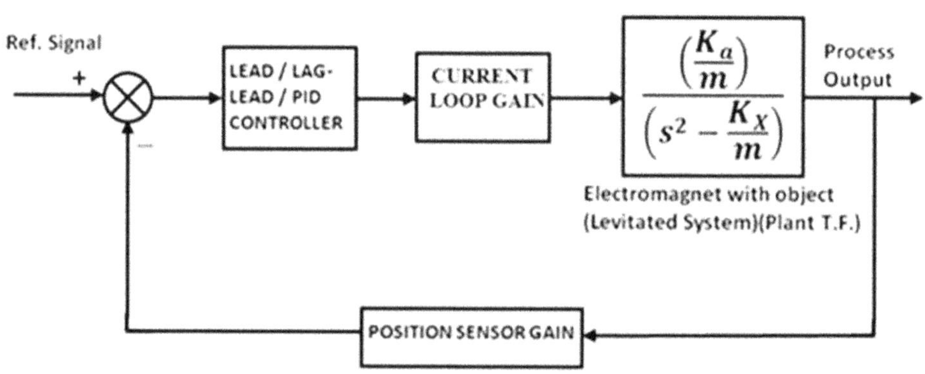

To design the position controller it is necessary to determine the model of the different subsystems of the position control loop. The transfer functions of the different blocks of the position control loop are determined in the following sub-sections.

Levitated System (Plant)

The dynamics of the levitated device is considered to be the plant transfer function of the system. Since the controlled current source is taken as the input of the actuator coil, the transfer function of the plant is given as

$$G_p(s) = \frac{\Delta X(s)}{\Delta I(s)} = \frac{\left(\frac{K_a}{m}\right)}{\left(s^2 - \frac{K_X}{m}\right)} \tag{4.28}$$

For the nominal operating air gap of 10 mm, and with m = 58 gm., K_a = 0.5358 (N/A) and K_X = 72.86 (N/m). After substituting all these values, the transfer function of the plant at 10mm air gap becomes

$$G_p(s) = \frac{\Delta X(s)}{\Delta I(s)} = \frac{9.237}{(s+35.44)(s-35.44)} \tag{4.29}$$

Current Loop

The details of the current loop with its different subsystems have already been described. The current loop may be represented by a block whose transfer function is found to be,

$$G_c(s) = \frac{K_{ch}\left(sK_p + K_i\right)}{\left\{s\left(R + sL\right) + \left(sK_p + K_i\right)K_{ch}K_c\right\}} \tag{4.30}$$

For the given parameters, the overall closed-loop transfer function of the current loop becomes,

$$G_c(s) = \frac{39.08s + 15930}{\{s(0.074s + 2.4) + (39.08s + 15910)\}} \tag{4.31}$$

REFERENCES

Banerjee, S., Prasad, D., & Pal, J. (2006). Large gap control in electromagnetic levitation. *ISA Transactions*, *45*(2), 215–224. doi:10.1016/S0019-0578(07)60191-8 PMID:16649567

Krishnan, R. (2003). *Electric motor drives: Modeling, analysis, and control.* Prentice Hall of India.

Moon, F. C. (1994). *Superconducting levitation, applications to bearings and magnetic transportation*. John Wiley & Sons.

Chapter 5
Design Fabrication and Testing of a Single Actuator DCALS (Part 2)

ABSTRACT

Modeling, design, fabrication, and testing of single magnet, single axis DC attraction type levitation system (DCALS) has been presented in this chapter. The proposed two loop scheme is more attractive from control point of view than the mostly reported single loop position control. The order of the levitated system (plant) has been reduced by one due to the use of a controlled current source as the excitation of the magnet-coil. The analysis and design of the position controller becomes easier after considering the reduced order plant transfer function. The controllers have been designed based on classical control theory utilizing root-locus and frequency domain technique. The realization of the controller also becomes simple due to the use of analog circuits.

DESIGN OF POSITION CONTROLLER

The transfer function of the plant $G_p(s)$ in Eqn. 5.20 shows that the system is unstable in open loop having a pole at $\left(+\sqrt{\frac{K_X}{m}}\right)$ in the right half's' plane. The root locus plot (Figure 1) of the closed loop system about an operating gap (10 mm) indicates that the system cannot be stabilized by simply adjusting

DOI: 10.4018/978-1-6684-7388-7.ch005

the amplifier gain. To make the levitation system stable, it is necessary to shift the root-locus to the left half of 's' plane.

For stabilization of unstable DCALS a cascade compensation control scheme utilizing PD, Lead, Lag-Lead, PID and combination of these controllers may be used. The reasons for its popularity can be attributed to simplicity, ease of design and robustness. The main aim of this part of work is to compare with the different designed controllers for the proposed designed DCALS at different operating points and to suggest a suitable classical controller that will provide fast response without any overshoots and steady-state error which is generally required for such type of critical systems.

Figure 1. Root locus plot of the uncompensated plant at 10mm gap

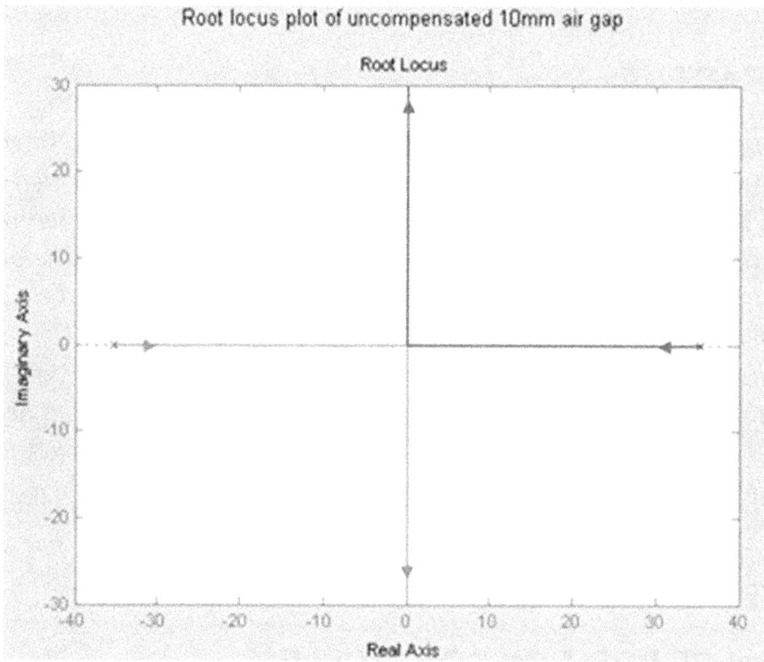

PD Controller

From the transfer-function as given in Eqn. 5.20, it is clear that the system is unstable in open loop and it can not be stabilized just increasing the gain of the system. Stabilization of this type of magnetically levitated system can easily be achieved by shifting the root-loci at the left-half of the s-plane (Figure 2). The addition of a zero to the open loop transfer function has the effect of pulling the root locus to the left, tending to make the system stable and to speed up the settling time of the response. Physically the addition of a zero in the feed forward transfer function means addition of proportional plus derivative control to the system. The transfer function of this controller is given as: $G_{pd}(s) = K_p + K_d s$. Here the value K_p is usually determined to satisfy the steady-state requirement. The corner frequency $\frac{K_p}{K_d}$ is chosen such that the phase lead occurs in the neighbourhood of the gain crossover frequency. Although the phase margin can be increased, the magnitude of the compensator gain continues to increase for the frequency range $\frac{K_p}{K_d} < \omega$. Such a continual increase in the magnitude of compensator gain is undesirable, since it amplifies high-frequency noise that is present in the electromagnetic levitation system. The step responses of the proposed DCALS at 10mm operating air-gap utilizing different controllers have been presented in Figure 11. It is seen that the compensated system with PD compensation is critically damped but it has large steady-state error. The compensated root-locus plot of the system operating at 10mm air-gap is shown in Figure 2.

Figure 2. Root locus plot of 10 mm plant with PD controller

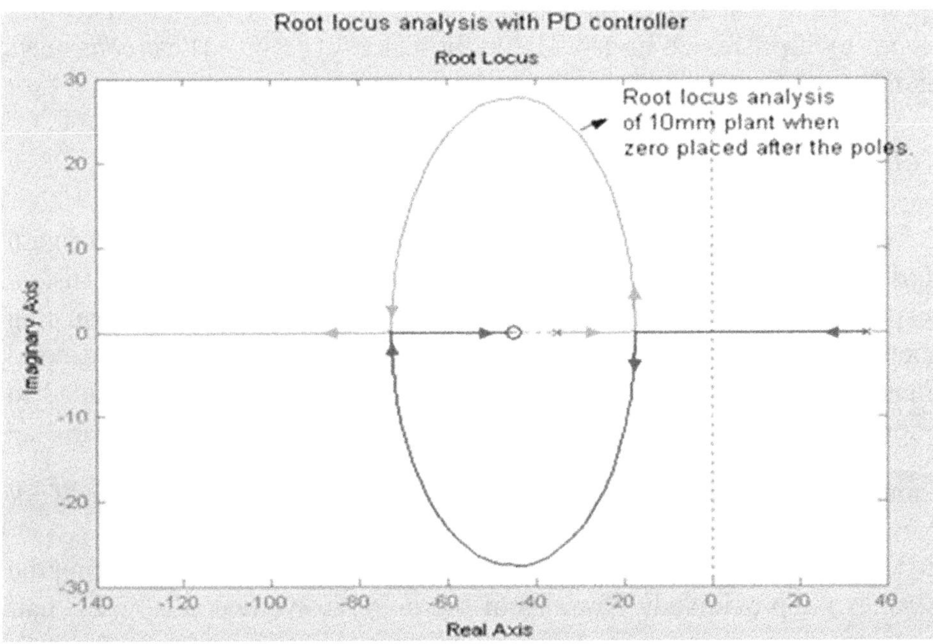

Lead Controller

Though PD controller theoretically stabilizes the closed loop DCALS, it will amplify the high frequency noise signal (mostly generated from high frequency chopping operation of switched mode power supply) during experimentation. Therefore, to reduce the effects of high frequency noise (due to EMI etc), a pole (p_c) is to be placed far away (approximately 10 times than the compensated zero location) along the negative real axis and this effectively produces a phase Lead controller whose transfer-function is given as: $G_{Lead}(s) = \frac{K(s + z_c)}{(s + p_c)}$ where K is the gain, z_c is the zero and p_c is the pole of the Lead controller (Clark, 1996; Ogata, 2000). Simulation studies are carried out to find a pole-zero and gain combination so that the system is stable as well as acceptable performance is obtained. The root locus plot of the stabilized system around 10 mm operating air-gap utilizing Lead compensation is shown in Figure 3. From Figure 4 and Table 1 it is seen that with the designed Lead controllers for different operating air-gaps the closed loop system becomes stable and satisfactory transient performances have been achieved but steady-state performance is poor since the Lead controller can not eliminate the offset. The DC gain of compensators keep on increasing since the inherent gain of plant is decreasing with the air-gap. To maintain same percent peak overshoot (nearly 16%) and steady-state error (13%), the system response becomes sluggish with increased air-gap. The performances of designed Lead controllers at different air-gap positions are shown in Figure 4.

Lead-Lag Controller

A Lead-Lag compensator combines the effects of a Lead compensator with those of a Lag compensator. The result is a system with improved transient response, stability and steady-state error. To implement a Lead-Lag compensator, first the Lead compensator is designed to achieve the desired transient response and stability, and then a Lag compensator is added to improve the steady-state response. The basic structure of the Lead-Lag compensator is $G_{Lead-Lag}(s) = \frac{K(s + z_{c1})(s + z_{c2})}{(s + p_{c1})(s + p_{c2})}$. In this work, the Lag network is designed by placing one pole (p_{c1}) very near to the origin and taking the zero (z_{c1}) approximately 10 times of the pole on –ve real axis at left half s-plane and Lead network is designed by placing one zero (z_{c2}) just at the adjacent left side of the levitated system's –ve pole and placing one pole (p_{c2}) approximately 20 times far away of that zero (z_{c2}) on –ve real axis at left half s-plane. From Figure 6 and Table 2, it is clear that introducing Lead-Lag network both the satisfactory transient and steady-state performance is achieved and it is aimed to obtain similar specifications (overshoot 16% and steady-state error 1%) irrespective of any operating air-gap. This idea gives an insight to develop a piecewise linear control algorithm for the possible enhancement of the operating air-gap of any DCALS.

With the introduction of Lag controller, the order of the system increases and hence the stability margin of the overall system reduces than with the Lead controller. But this controller reduces steady-state error drastically. The advantage of this scheme is that here both pole and zero can be suitably placed to get the desired performance which is not being possible in case of PID controller.

Figure 3. Root locus plot of 10mm plant with Lead controller

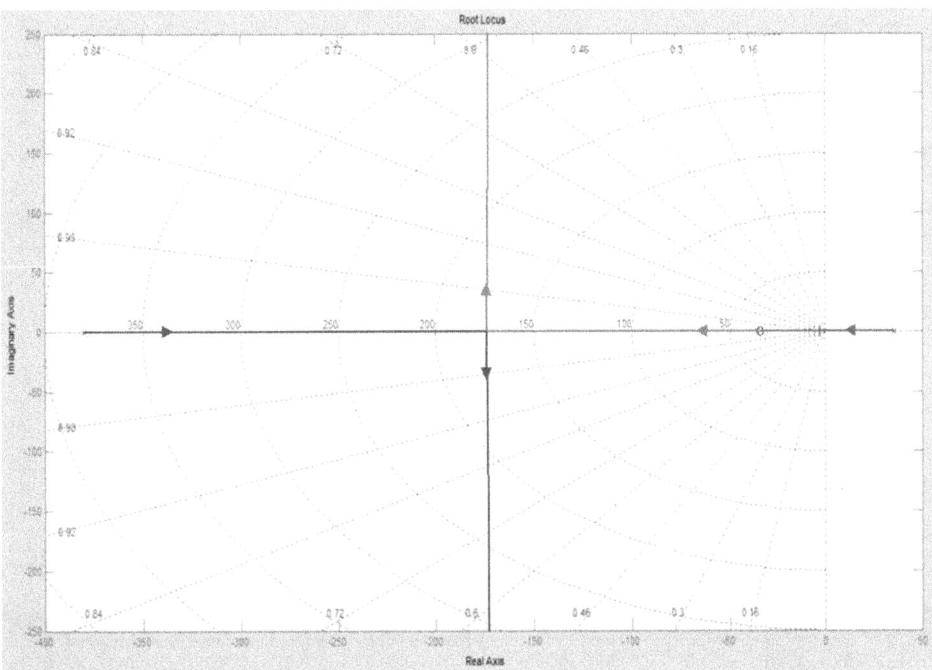

Table 1. Performances of Lead controllers at different operating positions

Air Gap (mm)	Transfer Function of the Maglev System	Transfer Function of Lead Controllers	Rise Time (sec)	P.O.*	Peak Time (sec)	Settling Time (sec)	Steady State Error
3mm	$\frac{43.26}{(s+83.47)(s-83.47)}$	$14.4\frac{(s+80)}{(s+813)}$	0.00220	16	0.00482	0.0113	0.13
5mm	$\frac{24.53}{(s+64.148)(s-64.148)}$	$14.7\frac{(s+60)}{(s+610)}$	0.00288	16	0.00625	0.0151	0.13
7mm	$\frac{15.368}{(s+46.62)(s-46.62)}$	$15\frac{(s+43)}{(s+485)}$	0.00359	16	0.00775	0.0190	0.12
10mm	$\frac{9.237}{(s+35.44)(s-35.44)}$	$15.3\frac{(s+31)}{(s+369)}$	0.00455	16	0.00975	0.025	0.12
12mm	$\frac{3.652}{(s+23.27)(s-23.27)}$	$15.5\frac{(s+19)}{(s+226)}$	0.00719	16	0.0157	0.041	0.13
15mm	$\frac{1.74}{(s+15.31)(s-15.31)}$	$16\frac{(s+13)}{(s+163)}$	0.01010	16	0.0226	0.0569	0.12
17mm	$\frac{0.2068}{(s+5.326)(s-5.326)}$	$16.8\frac{(s+5)}{(s+60)}$	0.02920	16	0.0642	0.152	0.11

* P.O.- Percentage Overshoot

Figure 4. Position responses with lead controllers at different operating points

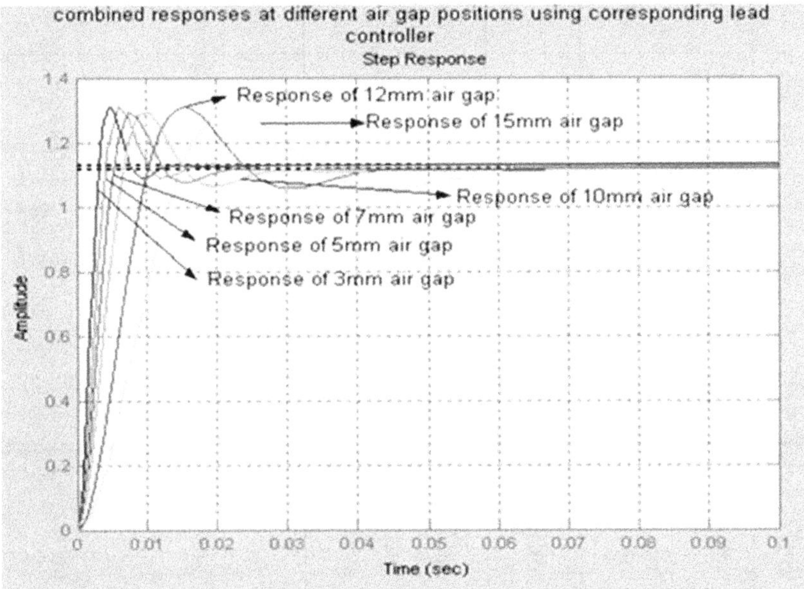

Figure 5. Root locus of the plant for 10mm gap with Lead- Lag controller

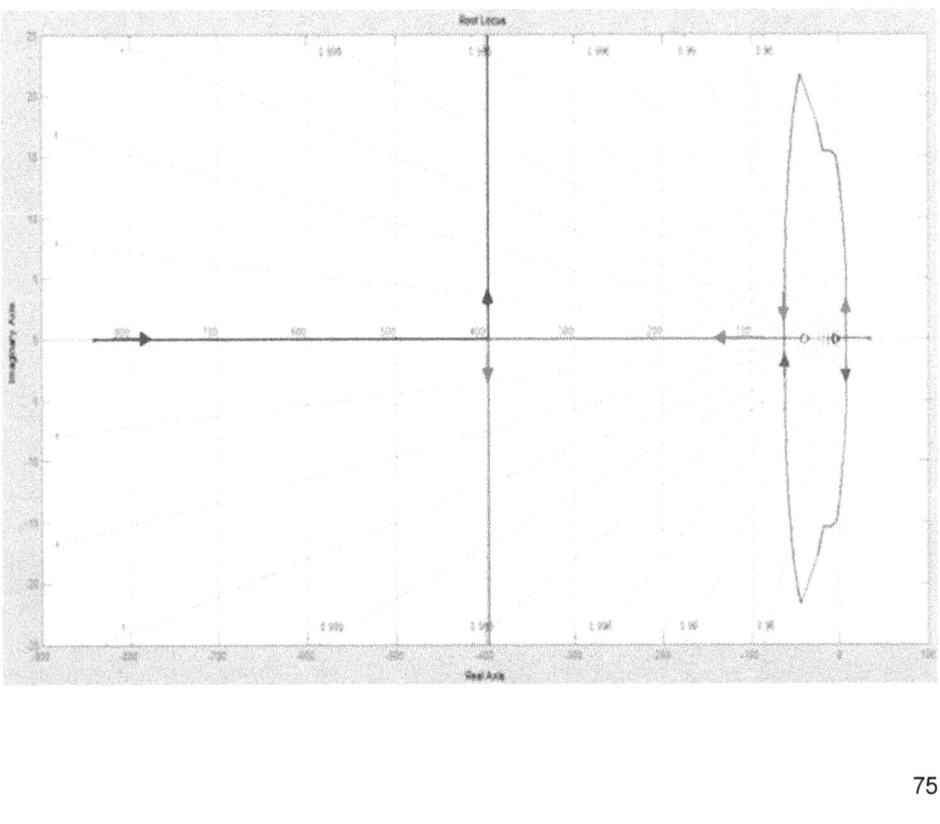

Table 2. Performances of Lead-Lag controllers at different operating positions

Air Gap (mm)	Transfer Function of the Maglev System	Transfer Function of Lead-Lag Controllers	Rise Time (sec)	P.O.	Peak Time (sec)	Settling Time (sec)	Steady State Error
3mm	$\frac{43.26}{(s+83.47)(s-83.47)}$	$22.9\frac{(s+5)(s+84)}{(s+0.5)(s+1610)}$	0.00238	16.1	0.00606	0.353	0.01
5mm	$\frac{24.53}{(s+64.148)(s-64.148)}$	$25\frac{(s+5)(s+65)}{(s+0.5)(s+1300)}$	0.00299	16.0	0.00781	0.356	0.01
7mm	$\frac{15.368}{(s+46.62)(s-46.62)}$	$25.5\frac{(s+5)(s+51)}{(s+0.5)(s+1170)}$	0.00392	16.2	0.01310	0.357	0.01
10mm	$\frac{9.237}{(s+35.44)(s-35.44)}$	$26.3\frac{(s+5)(s+40)}{(s+0.5)(s+840)}$	0.00483	16.0	0.01530	0.359	0.01
12mm	$\frac{3.652}{(s+23.27)(s-23.27)}$	$30\frac{(s+5)(s+26)}{(s+0.5)(s+600)}$	0.00708	16.2	0.02220	0.360	0.01
15mm	$\frac{1.74}{(s+15.31)(s-15.31)}$	$33\frac{(s+5)(s+18)}{(s+0.5)(s+450)}$	0.0104	16.2	0.03140	0.380	0.01
17mm	$\frac{0.2068}{(s+5.326)(s-5.326)}$	$54\frac{(s+5)(s+6)}{(s+0.5)(s+190)}$	0.02180	16.2	0.06270	0.372	0.01

Figure 6. Position responses with Lead- Lag controllers at different operating points

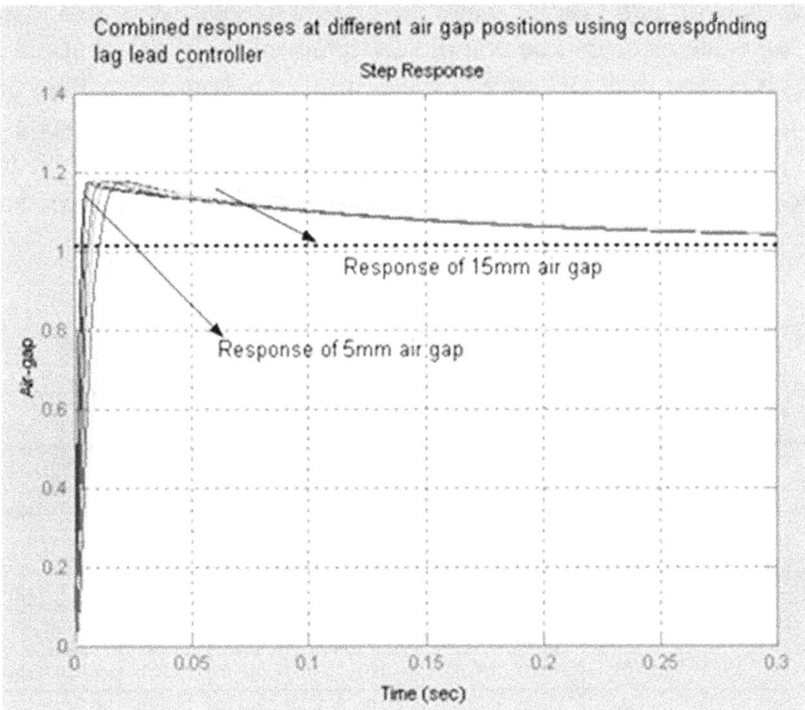

PID Controller

The transfer-function of the PID controller is given as:

$G_{PID}(s) = (K_p + sK_D + \frac{K_I}{s})$, where K_p, K_D and K_I are the proportional, derivative and integral gain respectively. Here the selection of three gains is such that the two zeros are to be placed at the negative real axis. So essentially the transfer-function becomes $G_{PID}(s) = \frac{K(s+z_{c1})(s+z_{c2})}{s}$; where proportional gain, $K_p = K(z_{c1}+z_{c2})$, derivative gain, $K_D=K$ and integral gain $K_I=K*z_{e1}*z_{e2}$. The choice of zeros of the PID controller is based on the location of the poles of the levitated system. Here, z_{e2} is placed very near (at the right side) to the stable pole of the plant and z_{e1} is placed near to the origin. After selecting the zeros at any operating point, a simulation is carried out to determine gain K enabling to achieve desired performance. Since the DCALS is inherently unstable system, the application of PID controller becomes critical due to having a pole at origin. The system can be stabilized for a smaller value of gain (Table 3) and the margin of stability is also less. But the steady-state error can be completely removed with PID control. From Table 3 it is seen that, DC gain of the controller becomes less than that of Lead- Lag compensator (Table 2). Figure 7 shows the position responses of the levitated system with PID controller at different operating positions. Root-locus plot of the overall closed-loop system (Figure 8) shows that the system becomes conditionally stable.

Table 3. Performances of PID controllers at different operating positions

Air Gap	Transfer Function of the Maglev System	Transfer Function of PID Controllers	Rise Time (sec)	P.O.	Peak Time (sec)	Settling Time (sec)	Steady State Error
3mm	$\frac{43.26}{(s+83.47)(s-83.47)}$	$\frac{0.0147(s+5)(s+72)}{s}$	0.00266	16	0.0207	0.386	0
5mm	$\frac{24.53}{(s+64.148)(s-64.148)}$	$\frac{0.0194(s+5)(s+58)}{s}$	0.00347	16	0.0208	0.386	0
7mm	$\frac{15.368}{(s+46.62)(s-46.62)}$	$\frac{0.0224(s+5)(s+45)}{s}$	0.00467	16	0.0214	0.382	0
10mm	$\frac{9.237}{(s+35.44)(s-35.44)}$	$\frac{0.0285(s+5)(s+35)}{s}$	0.00607	16	0.0254	0.383	0
12mm	$\frac{3.652}{(s+23.27)(s-23.27)}$	$\frac{0.0485(s+5)(s+23)}{s}$	0.00897	16	0.0361	0.392	0
15mm	$\frac{1.74}{(s+15.31)(s-15.31)}$	$\frac{0.0695(s+5.5)(s+14)}{s}$	0.0132	16	0.0516	0.391	0
17mm	$\frac{0.2068}{(s+5.326)(s-5.326)}$	$\frac{0.264(s+4)(s+7.5)}{s}$	0.0274	16	0.0876	0.388	0

Figure 7. Position responses with PID controllers at different operating points

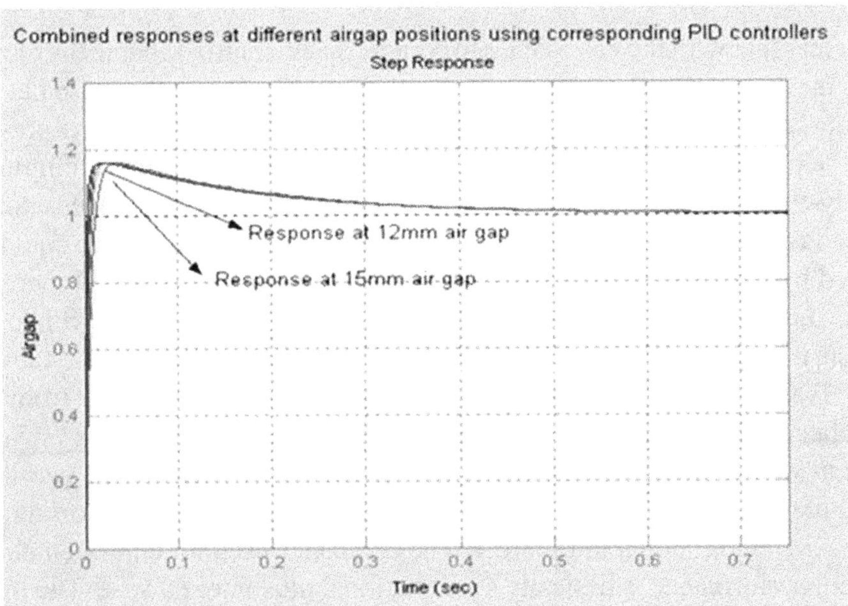

Figure 8. Root locus plot of 10mm plant with PID controller

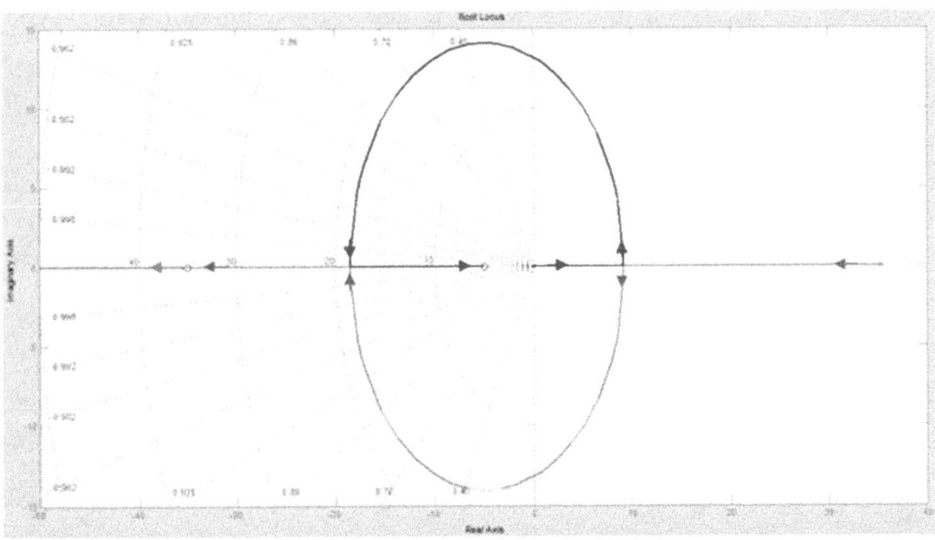

PI (Outer Loop) Plus Lead (Inner Loop) Controller

It is clear that with the cascade Lead compensated control scheme the closed loop maglev system is stable and has satisfactory transient response but the system steady-state accuracy is not good. Steady-state error is always present for a step input, because for this system Lead compensation cannot eliminate steady-state error for a step input. Introducing another Lag compensator in cascade cannot reduce the steady-state error to zero. This problem could be solved by introducing an integrator in the control loop, but it will increase the difficulty of designing a stabilizing controller for the system. Since the system is unstable, the single controller used so far cannot satisfy both stability as well as performance requirements (Clark, 1996), which is clear from the simulated results (Figure 4). To enhance the system performance (steady-state as well as transient) another loop must be introduced into the two-loop structure of DCALS presented in Figure 4. The modified schematic diagram of the system is shown in Figure 9. The controller used for enhancement of system performance is basically a proportional plus integral type. The inner Lead controller will stabilize the overall closed loop system and the outer PI control will improve the performance of the system.

From Figure 10 and Figure 11, it is observed that the response has no overshoot and steady-state error which is generally required for such type of critical system.

Figure 9. Modified closed loop maglev system utilizing outer PI controller

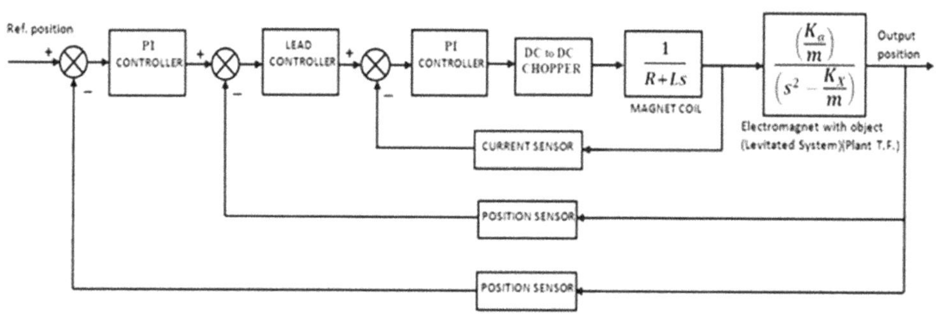

The performance of the controllers at four different air gap positions is shown in Figure 10. It appears that all the controllers are satisfying almost same performance at the respective air gap position.

Figure 10. Position responses with outer PI controller while switching from one gap to other

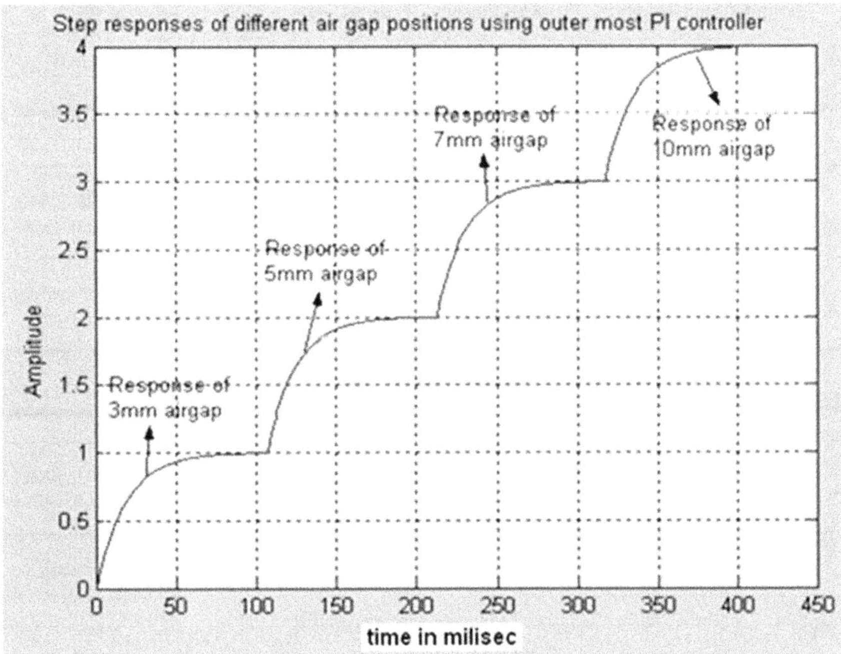

Figure 11. Position responses due to step input for the different controllers at 10 mm air-gap position

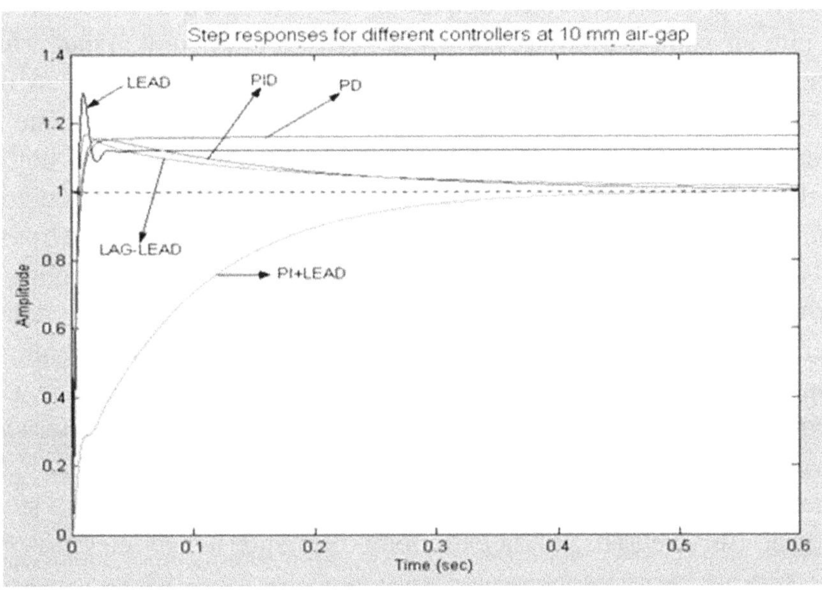

Table 4. Performances of PI plus Lead controller at different operating positions

Air Gap (mm)	Transfer Function of the Plants	Transfer Function of Outer Most PI Controller	Rise Time (sec)	Settling Time (sec)	GM (PCF) & PM (GCF)
3mm	$\frac{43.26}{(s+83.47)(s-83.47)}$	$\frac{0.0315(s+85)}{s}$	0.561	0.99	27.1db(460rad/sec) 91.4deg(4.02rad/sec)
5mm	$\frac{24.53}{(s+64.148)(s-64.148)}$	$\frac{0.036(s+80)}{s}$	0.562	0.99	27.5db(428rad/sec) 91.6deg(4.04rad/sec)
7mm	$\frac{15.368}{(s+46.62)(s-46.62)}$	$\frac{0.043(s+70)}{s}$	0.571	1.00	27.7db(389rad/sec) 91.6deg(3.99rad/sec)
10mm	$\frac{9.237}{(s+35.44)(s-35.44)}$	$\frac{0.064(s+50)}{s}$	0.571	1.00	27.9db(357rad/sec) 93deg(4.07rad/sec)
12mm	$\frac{3.652}{(s+23.27)(s-23.27)}$	$\frac{0.082(s+40)}{s}$	0.572	1.01	30.7db(292rad/sec) 93.5deg(4.07rad/sec)
15mm	$\frac{1.74}{(s+15.31)(s-15.31)}$	$\frac{0.104(s+32)}{s}$	0.575	1.00	30.8db(268rad/sec) 93.6deg(4.00rad/sec)
17mm	$\frac{0.2068}{(s+5.326)(s-5.326)}$	$\frac{0.17(s+23)}{s}$	0.575	1.01	34.6db(236rad/sec) 97.1deg(4.31rad/sec)

Different position responses due to step input for the different controllers at 10 mm air-gap position are shown in Figure 11. Though apparently PD compensator provides better transient response, practically due to noise problem, a modified PD controller will be required after adding one high frequency pole that essentially becomes a Lead controller. But both the controller can not eliminate the steady-state error. PID and Lead-Lag controller is always better than Lead or Lag controller for any system both from transient and steady-state accuracy point of view. But due to use of these controllers the order of the system becomes higher and the margin of stability reduces. PID controller demonstrates better performance than Lead-Lag controller both in transient and steady-state but the margin of stability is less than Lead-Lag controller. In the present case the designed Lead-Lag controller shows faster response than the PID controller. The PI plus Lead controller exhibit a critically damped response without having no overshoots and zero steady-state error. But due to use of an extra loop the response becomes slower.

CONTROL CIRCUIT HARDWARE

The proposed single actuator based system (Figure 1 in Chapter 4) has been tested with the designed classical controllers. Figure 12 shows the overall control circuit with cascade Lead compensation for the proposed levitation system. The different parts of the control circuit are generation of reference command (position gap) signal, position controller, current controller, PWM generation for the chopper switch and driver circuit for the MOSFET switch used in the chopper circuit. The individual parts of the circuit are marked in the overall control circuit diagram (Figure 12).

Figure 12. Overall control circuit diagram

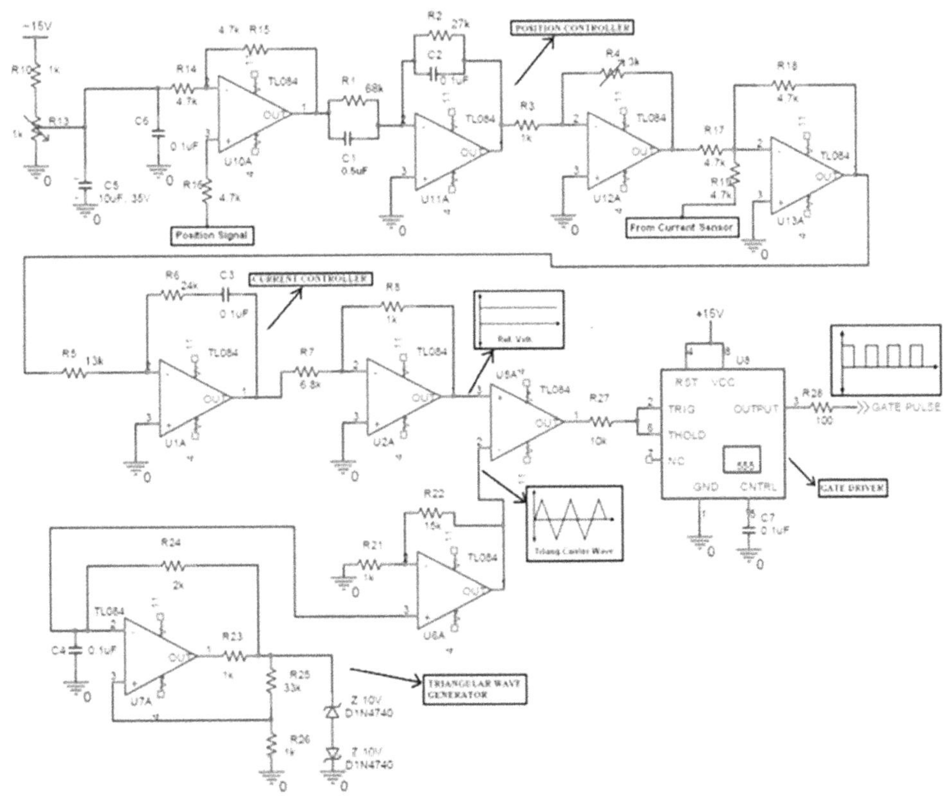

REALIZATION OF POSITION CONTROLLER

The position controller used in this work is basically a phase Lead compensator as described in earlier section. The general form of transfer function of a Lead controller is shown in Figure 13. The overall transfer-function of the network is given as follows:

$$\text{or, } \frac{V_o(s)}{V_i(s)} = \frac{R_4 C_1}{R_3 C_2} \left(\frac{s + \frac{1}{R_1 C_1}}{s + \frac{1}{R_2 C_2}} \right) \tag{4.32}$$

From Table 4 the transfer of designed Lead controller at 10mm air-gap is

$$G_{Lead}(s) = 15.3 \frac{(s+31)}{(s+369)} \tag{4.33}$$

Now comparing Eqn.4.32 and Eqn.4.33,

$$\frac{1}{R_1 C_1} = 31$$

$$\frac{1}{R_2 C_2} = 369$$

Let us assume, $C_1 = 0.5\mu F$, $C_2 = 0.1\mu F$ and hence R_1 comes out to be 64.5 KΩ and $R_2 = 27.1$ KΩ. For the present case nearest standard values of resistors and capacitors have been considered.

REALIZATION OF CURRENT CONTROLLER

In the current control loop the actual current is compared with the reference current. The actual current is sensed by LEM make Hall-effect current sensor and after scaling (1 amp = 1 volt) it is fed to a differential amplifier circuit.

The other input to the differential amplifier is the reference current signal generated from the position controller output. The current error signal output from the differential amplifier is processed through a PI controller.

Figure 13. Schematic diagram of Lead controller circuit

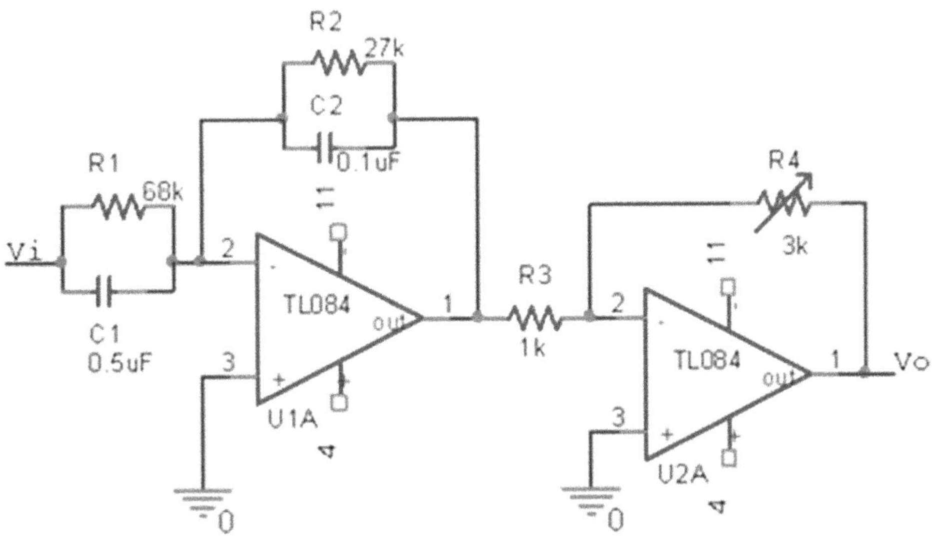

The circuit of the current controller as shown in Figure 14 and is given as

$$G_{cc}(s) = \frac{\left(R_6 + \frac{1}{sC_3}\right)}{R_5} = \left(\frac{R_6}{R_5} + \frac{1}{sR_5C_3}\right) \tag{4.34}$$

Comparing with the standard form of PI controller, the proportional gain, $K_p = \frac{R_6}{R_5}$ and the integral gain, $K_i = \frac{1}{R_5C_3}$;

Therefore, as per designed current (PI) controller (Section 4.4.2.1)

$$K_p = \frac{R_6}{R_5} = 1.711 \text{ and } K_i = \frac{1}{R_5C_3} = 707.3\,,$$

Figure 14. Schematic diagram of PI controller circuit

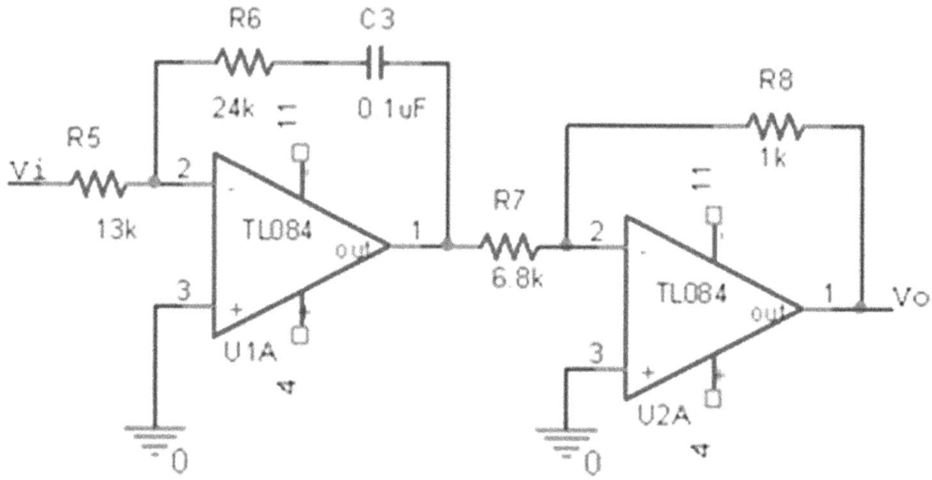

DETAILS OF THE TRIANGULAR WAVE GENERATION

An increasing (positive direction) output voltage from current controller will mean that reference current demand is more and actual current to the magnet (sensed by Hall sensor) is less. Reverse will be true when the current controller output is decreasing. Thus to meet the magnet current demand, the chopper duty cycle needs to be changed according to the current sensor output. The low frequency current sensor output (modulating signal) is compared with a high frequency triangular wave and the comparator output decides the duty ratio of the DC to DC chopper circuit. The circuit diagram of the triangular wave generator is given in Figure 15. For triangular wave generation, an op-amp is used in astable multivibrator mode and another op-amp is put in cascade as a non-inverting buffer amplifier. The output from first op-amp is a square wave that is clipped to desired voltage of ±10 volts by means of two zener diodes connected in back to back fashion. A small part of the output voltage (across zener) is fed to the non-inverting pin of the op-amp as the reference voltage. Simultaneously a 0.1 μF capacitor connected between the inverting input of the op-amp and ground is charged from the op-amp output through a charging resistor of $2k\Omega$. In the low voltage range the capacitor charges more or less linearly. When the voltage across this capacitor exceeds the small reference voltage at the non-inverting input, the op-amp output changes polarity. Now the polarity of reference voltage reverses and the capacitor starts charging linearly in the reverse direction. The sequence repeats at a frequency decided by the R-C time constant of the multi-vibrator and the magnitude of the small reference voltage at the non-inverting input of op-amp. A square wave is produced by the op-amp and a small triangular wave is produced across the capacitor. The buffer op-amp gain and the multi-vibrator parameters are chosen to produce a triangular wave of ±5 volts peak. This triangular wave is eventually used to generate the required duty ratio for the chopper switch.

Figure 15. Hardware circuit for triangular wave generation

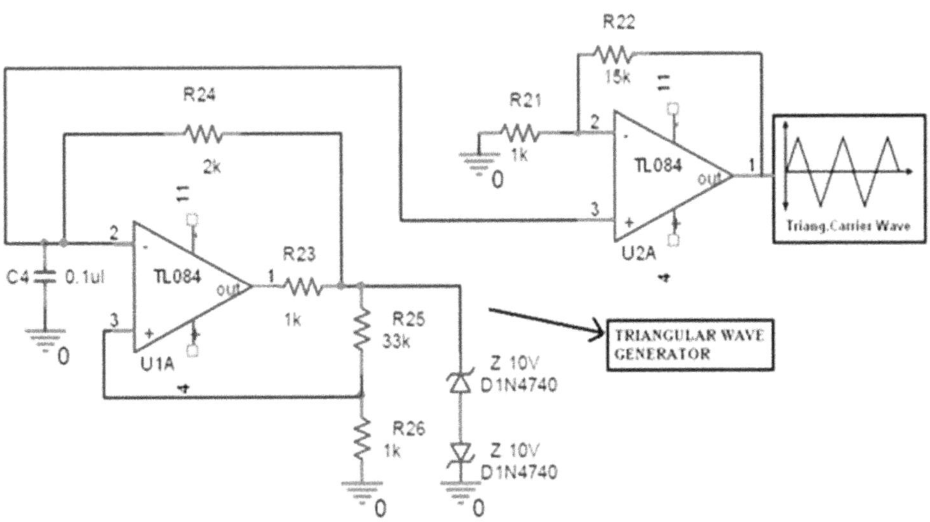

GATE DRIVE CIRCUIT

Comparison of triangular carrier wave and low frequency modulating wave (output of current controller) decides the duty ratio of the DC to DC chopper. Since in this work a single switch based DC to DC chopper (Figure 5 in Chapter 3) is used, separate isolator circuit may be avoided and the output of comparator (Figure 16) is directly connected to the astable multivibrator for amplification. The comparator (realized by an op-amp TL084) compares the triangular wave and output of the current controller. The duty ratio of the MOSFET switch varies as the rod moves up and down within the electromagnetic field. When the rod moves upwards, (beyond reference gap) the duty ratio of the MOSFET gate pulse is decreased, consequently the magnet current decreases and causes it to go down and vice-versa. To facilitate proper duty ratio control of the MOSFET switch, a triangular wave of 5 kHz is produced (described before). The triangular waveform is compared with the PI controller output of the current loop. The comparator generates the desired PWM waveform to be used for driving the MOSFET switch after amplification. The 'ON' time of the switch is decided by the duration for which the PI controller output remains higher than the triangular wave. Similarly, during 'OFF' period of the switch, the PI controller output becomes less than the triangular wave. The output of the comparator is not sufficient to drive the gate of the MOSFET switch and it is fed to an inverting buffer circuit using 555-IC. The output of 555 inverting buffer is connected to the gate of the MOSFET through 100Ω resistance.

Figure 16. Schematic diagram of gate drive circuit

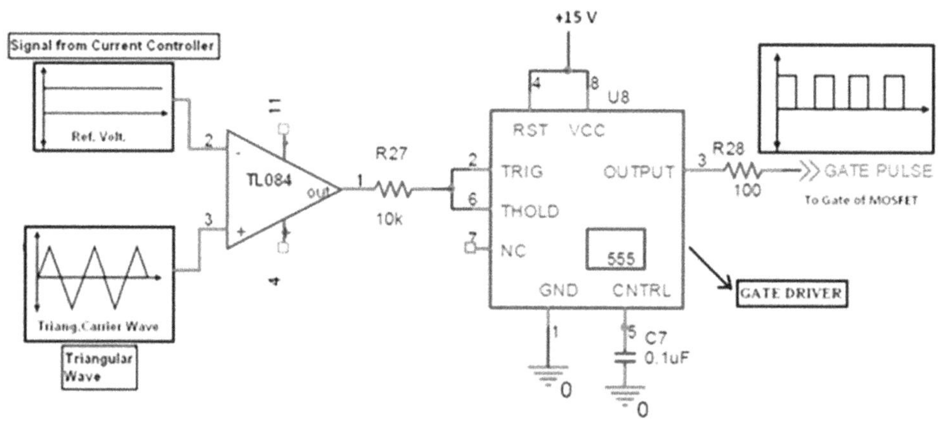

POWER AMPLIFIER

In this work a simple single switch based switched mode converter circuit has been utilized as the power circuit. The circuit diagram of the proposed power circuit has been given in Figure 5 (Chapter 3). The description and analysis of the circuit has already been presented in Chapter 3.

Figure 17. Simplified chopper circuit for single actuator based levitation system

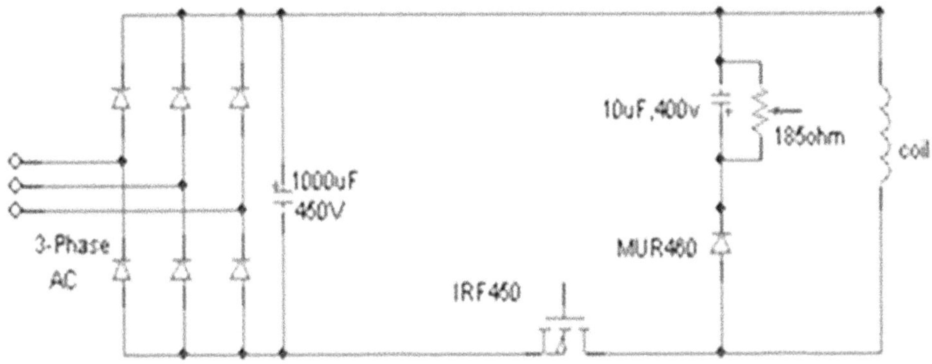

EXPERIMENTAL RESULTS

Figure 18 shows coil-voltage and coil-current during stable levitation of the ferromagnetic cylindrical rod. It is observed that when the switch is turned on, full supply voltage appears across the coil and the current in the coil increases and when the switch is turned off, capacitor voltage with negative polarity appears across magnet-coil, current in the coil decreases.

Figure 18. Coil-voltage (Ch. 2) and coil-current (Ch. 1) during stable levitation of cylindrical rod

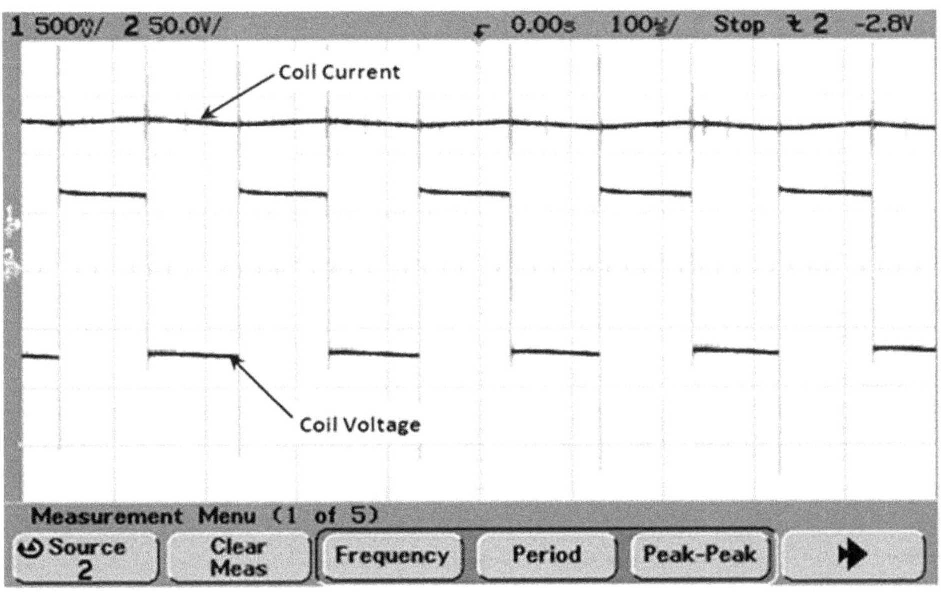

Figure 19. Capacitor voltage (Ch. 1), coil voltage (Ch. 2) during stable levitation of cylindrical rod

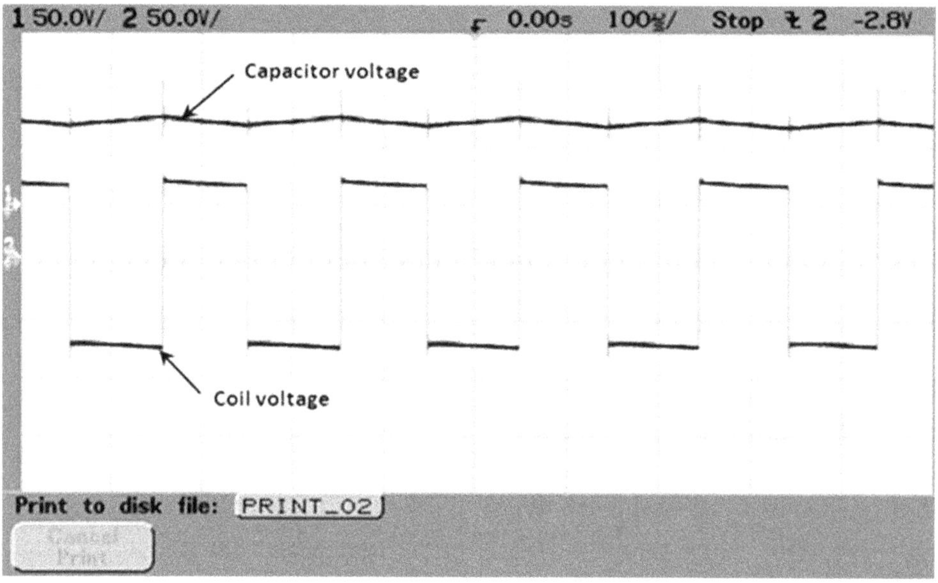

Figure 19 shows the charging and discharging voltage of capacitor (CH-1) and coil voltage during stable levitation of the object. During ON time the energy dump capacitor (10 μF, 400V) that holds some charge due to previous operation, discharges partially through the rheostat (185Ω). During OFF period of the switch, the coil currents find a path through fast-recovery power diodes to the resistor capacitor combination. The capacitor voltage effectively appears across the coils with a negative polarity (reverse of the polarity during ON duration). The coil voltage is made equal positive and negative supply voltage with the predefined circuit parameters (resistor and capacitor).

The cylindrical rod has been successfully suspended under the E-core electromagnet. The photograph of the complete hardware setup is shown in Figure 20 during levitated condition with around 10 mm distance between magnet face and object. The side view of levitation is shown in Figure 21.

Figure 22 to Figure 24 shows the position responses of the levitated object at nominal operating gap position (10 mm) utilizing different controllers. Here, the controller designed for a fixed air-gap is fed with a steady gap command superimposed with a small repetitive step voltage.

It is seen that with the use of Lead controller (Figure 22) the response is faster but having more overshoots and undershoots before reaching to final steady-state point. The steady-state performance is also poor since lead controller cannot eliminate steady-state error.

Figure 24 shows the experimental dynamic position response with PID controller at the nominal operating condition. As expected both the transient and steady-state performances have been improved than the response with Lead controller.

As expected from simulation results (Figure 10 and Figure 11), the experimental result (Figure 24) for the case of two loop position controller (using PI plus Lead controller) the response has no overshoot and steady-state error. But the response becomes sluggish due to use of one more loop.

Figure 20. Complete hardware set-up for single actuator based DC attraction type levitation system

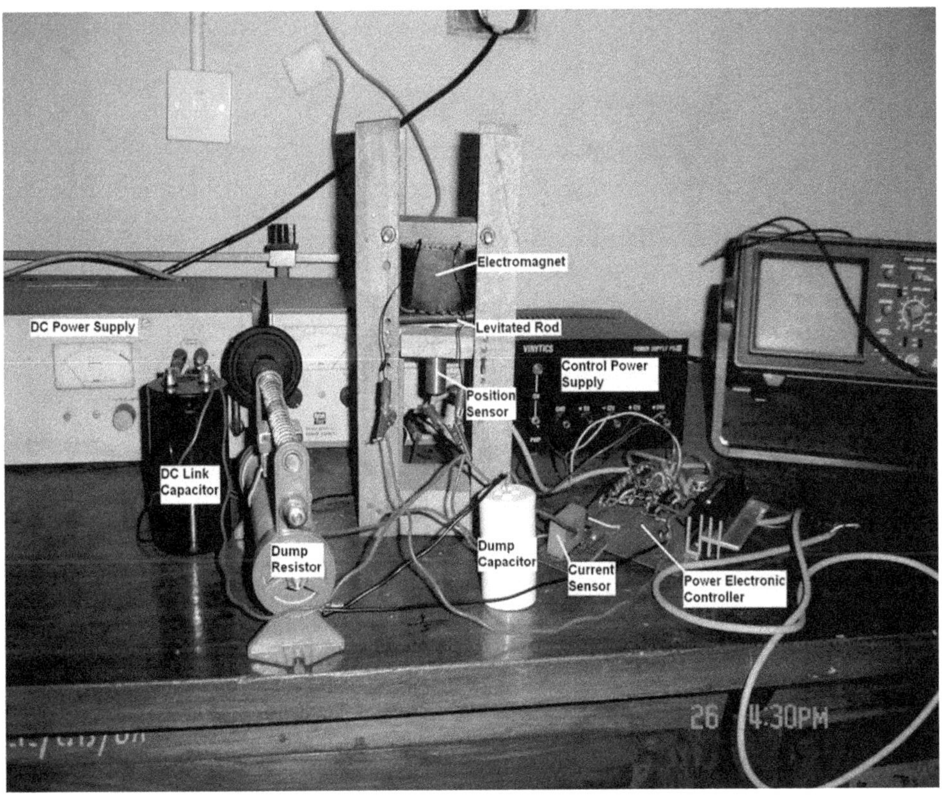

Figure 21. Side view of stable levitation of rod at nominal operating point

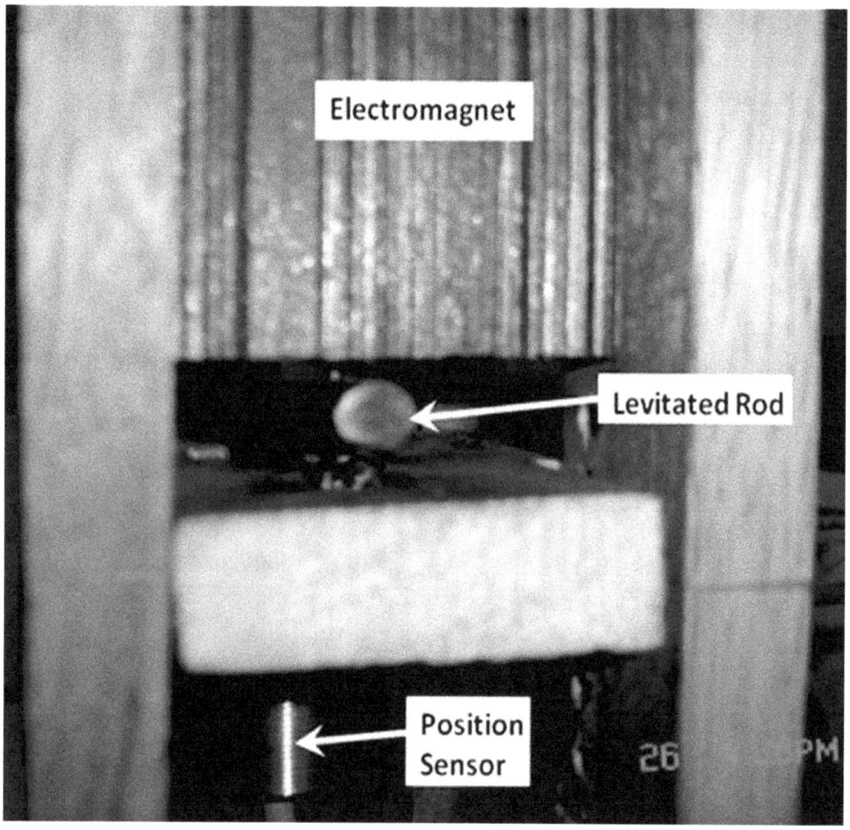

Figure 22. Position response with Lead controller at 10 mm operating gap

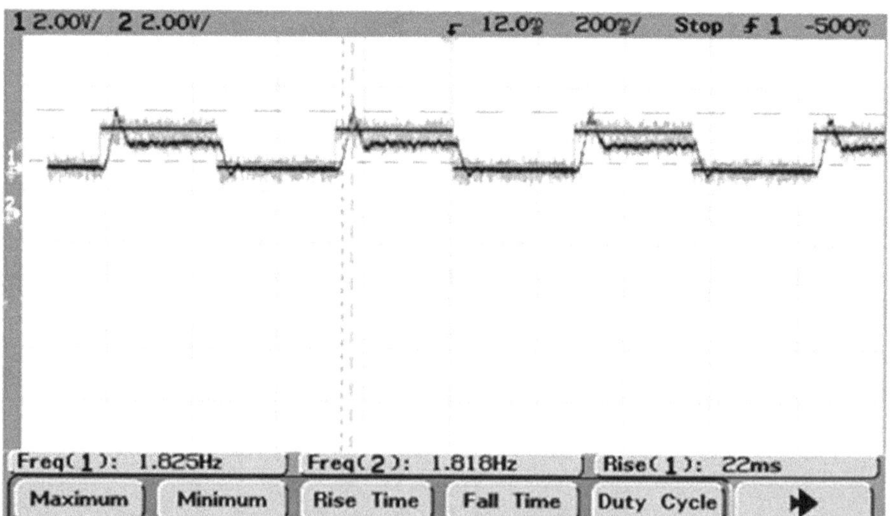

Figure 23. Position response with PID controller at 10 mm operating gap

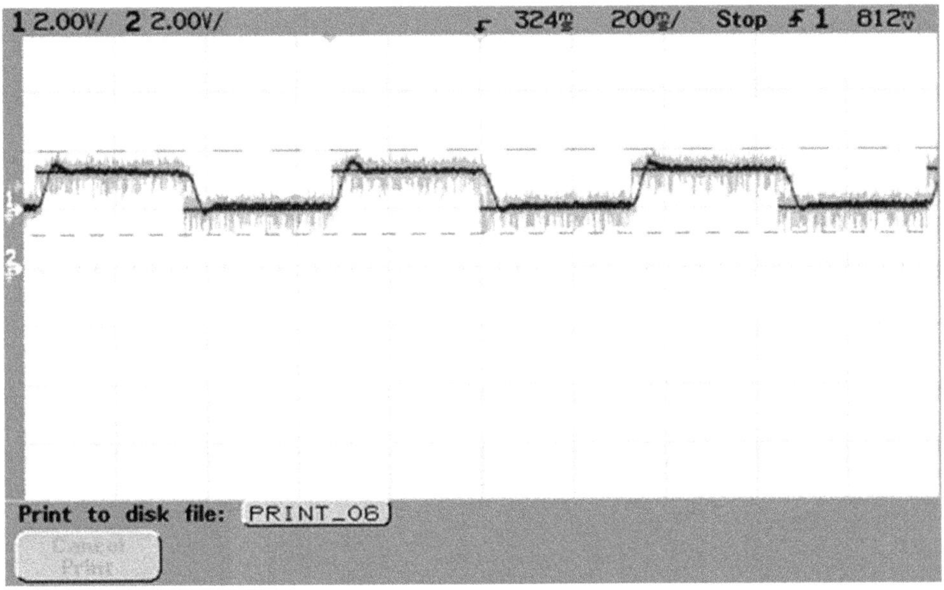

Figure 24. Position response with outer loop PI controller at 10 mm operating gap

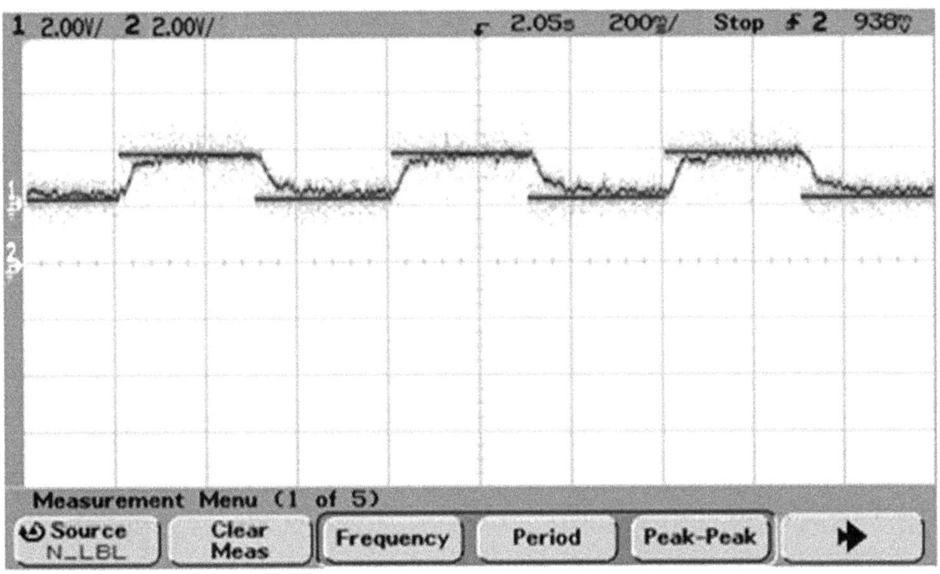

REFERENCES

Clark, R. N. (1996). *Control system dynamics*. doi:10.1017/CBO9781139163873

Ogata, K. (2000). *Modern control engineering* (3rd ed.). Prentice Hall of India.

Chapter 6
Effect of Different Parameters on the Controller Performance for a DC Attraction Type Levitation System

ABSTRACT

The basic parameters for electromagnetic levitation systems are inductance and resistance of the actuator, input DC link voltage, mass of the payload (object), air-gap between the pole face of the electromagnet and ferromagnetic guide-way or object, etc. All these parameters are supposed to change in real-life situations. DCALS is inherently unstable and strongly nonlinear in nature. In most cases, a classical controller has been designed for the linear model around an operating point for maintaining overall closed-loop stability. So, the performances of the linear controller with the change in different parameters have been studied. The parameters of the designed controller also vary during experimentation. So, a sensitivity study of the controller for a DCALS is also important.

INTRODUCTION

The basic parameters for electromagnetic levitation systems are inductance and resistance of the actuator, input DC link voltage, mass of the payload (object), air-gap between the pole face of the electromagnet and ferromagnetic

DOI: 10.4018/978-1-6684-7388-7.ch006

guide-way or object etc. All these parameters are supposed to get change in real life situation. DCALS is inherently unstable and strongly nonlinear in nature. In most cases a classical controller has been designed for the linear model around an operating point for maintaining overall closed-loop stability. So the performances of the linear controller with the change in different parameters have been studied (Banerjee & Bhaduri, 2008). The parameters of the designed controller also get varied during experimentation. So a sensitivity study of the controller for a DCALS is also important.

The proposed DCALS has been described already in previous chapters and the basic block diagram of the overall control system is shown in Figure 4 in Chapter 4. In this system a cylindrical rod has been arranged to suspend under E-core electromagnet at different operating air-gaps. The basic parameters for the system have been given in Table 1 in Chapter 4. In this work, a cascade compensation control scheme utilizing inner current loop and outer position loop has been designed and implemented for stabilization of a single magnet based single axis levitation system. The prototype has been successfully tested and stable levitation has been demonstrated at the desired operating gap. The dynamic performance of the controller has been tested by applying different disturbance inputs with increasing magnitudes and frequencies.

As described in Chapter 4 & 5, the transfer-function of the levitated system (given in Eqn.4.9) at 10 mm operating air-gap has been determined experimentally and is given by

$$G_p(s) = \frac{9.237}{(s+35.44)(s-35.44)} \tag{6.1}$$

For the operating air-gap of 10 mm, the plant $G_p(s)$ has one stable pole at s = -35.44 and an unstable pole at s = 35.44. The objective is to design a phase Lead compensator (Ogata, 2000) so that the overall closed loop system becomes stable. The pole and zero are to be placed on the negative real axis and there are many possible pole-zero locations for stabilizing the system. The phase-Lead compensator design procedure in this case is to place the zero of the compensator between 0 and –35.44 on the real axis of the s-plane, while the pole of the compensator is placed about 12 times to the left of the zero position. Simulation studies are carried out to find a pole-zero and gain combination so that the system is stable as well as acceptable performance is obtained. The transfer function of the designed Lead controller for an operating air-gap of 10 mm is given by

$$G_{Lead} = \frac{K(s+z_c)}{(s+p_c)} = \frac{15.3(s+31)}{(s+369)} \quad (6.2)$$

EFFECT OF CHANGE IN AIR-GAP

For any electromagnetic levitation system the air-gap between the pole-face of the magnet and ferromagnetic object may change due to many reasons such as change in pay-load, change in input DC-link voltage, change in sudden disturbance etc. So the performance of the linear controller (designed for any operating air-gap) at the changing operating condition is important. Here, the controller is designed for 10 mm operating air-gap for maintaining overall closed-loop stability and to satisfy some specified performances. Due to change in operating condition (both in higher and lower operating air-gap) the performance of the controller deteriorates. For the lower operating air-gap the inherent gain of the system has been increased (Table 2 of Chapter 4) making the system little under-damped and the dynamic response becomes faster (below Figure 11). But for higher operating air-gap due to low inherent system gain, the damping has been increased reducing the overshoot and making the system sluggish. The steady-state error has been increased slightly. As expected, due to inherent nonlinear characteristics of the system, the uncertainty in the nominal linearized model increases as one moves away from the nominal operating point of 10 mm.

EFFECT OF CHANGE IN MASS (PAYLOAD)

For an electromagnetic levitation system, the effect of change in payload is significant irrespective of any application. In some applications like maglev train, the change of payload is prominent and so the effect of change in mass on the controller performance is important.

The simplified closed loop maglev system with Lead controller is shown in Figure 2. From the figure the characteristic equation of the system is given by Eqn.6.3,

$$1 + \frac{K(s+z_c)}{(s+p_c)} \frac{K_a}{(ms^2 - K_X)} = 0 \quad (6.3)$$

Figure 1. The performance of the controller with the change of operating air-gap

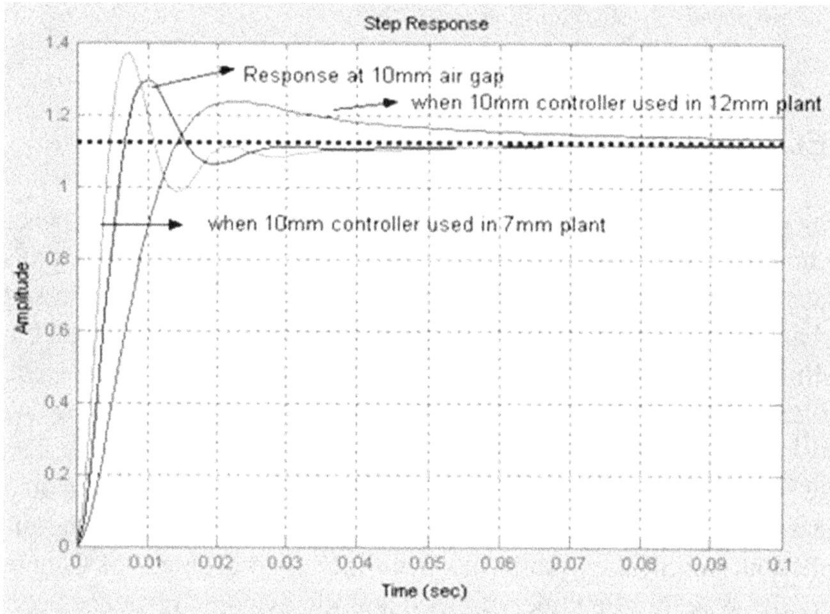

Figure 2. Simplified closed loop maglev system

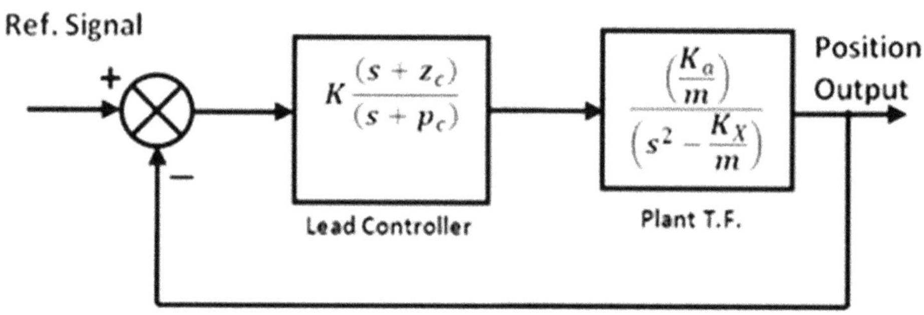

Neglecting the effect of high frequency pole of the Lead controller on the transient performance, Eqn.6.3 becomes,

$$\Rightarrow ms^2 - K_X + KK_a(s + z_c) = 0$$

$$\Rightarrow s^2 + \frac{KK_a}{m}s + \frac{(KK_a z_c - K_X)}{m} = 0 \tag{6.4}$$

So, the damping ratio (ζ) of the system may be written as follows

$$\zeta = \frac{KK_a}{2\sqrt{m}\sqrt{(KK_a z_c - K_X)}} \tag{6.5}$$

Here, K is the overall controller gain of the system, K_a and K_X are the two force constants, m is mass of the levitated object, z_c is the zero location of the Lead controller.

From Eqn.6.5, if the mass of the object is increased (keeping other parameters constant), then the damping ratio will decrease and vice-versa. From Figure 2, with the increase of the mass the system open loop poles are coming closer to the imaginary axis thereby reducing the overall closed-loop stability. From the simulated response shown in Figure 3, it is clear that with the increase in mass the system exhibits more oscillation with faster response and vice-versa.

Figure 3. The performance of the controller with the change in mass of object

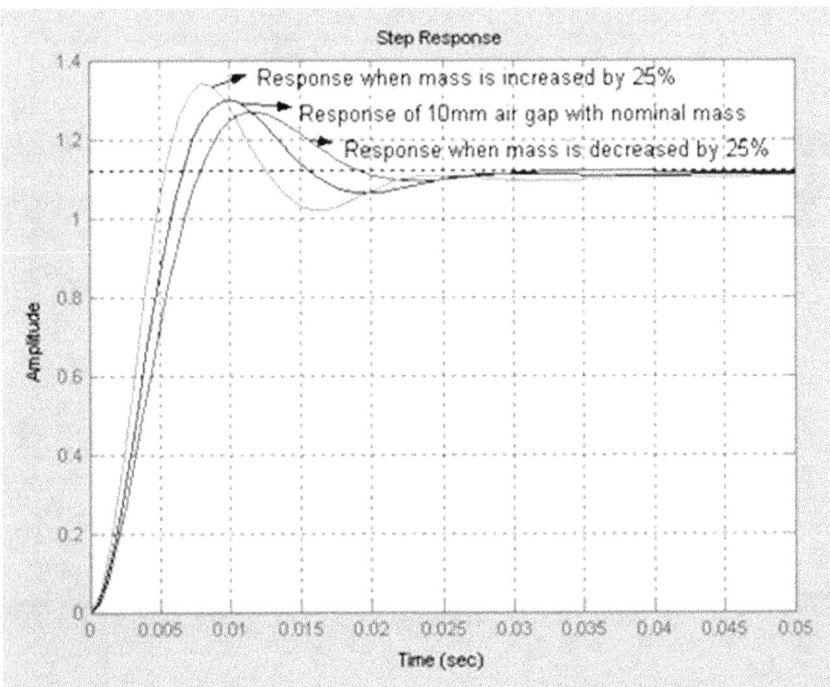

EFFECT OF CHANGE IN COIL RESISTANCE

The coil resistance may be increased due to the change in operating temperature for prolong operation of the levitated system. The heating of electromagnet depends on mass of object, duration of operation, operating air-gap, coil gauge etc. It has been observed that the effect of coil resistance on the dynamic performance of controller is less. In the designed two-loop control (Figure 4, Chapter 4), the innermost faster current loop takes care of magnet time constant which is decoupled from maglev system. So the coil resistance only appears in the innermost current loop. As the inner current loop is faster than the outer position loop, the inner loop dynamics may be neglected while designing the outer loop position controller. From the frequency response (Fig.6.4 & Fig.6.5), it is seen that there is no such appreciable change in GM and PM even with the 50% enhancement in coil resistance (which is not practically feasible also).

Figure 4. Frequency response at 10 mm air-gap with nominal coil resistance

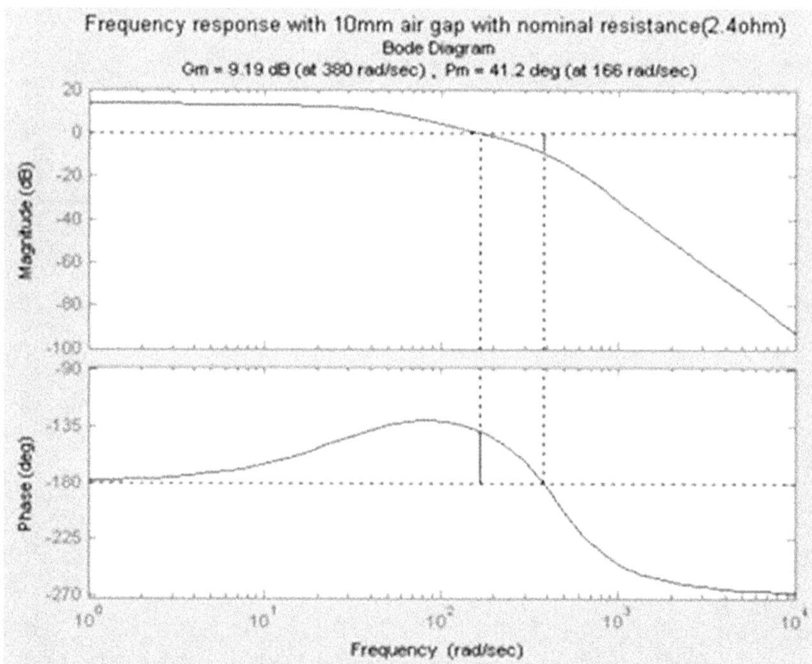

Figure 5. Frequency response when resistance is increased by 50%

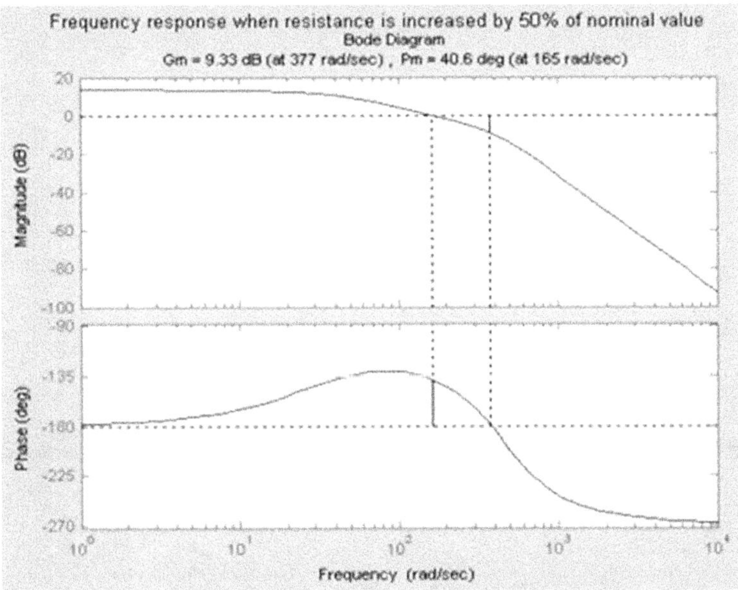

EFFECT OF CHANGE IN COIL- INDUCTANCE

From Eqn.4.6 to Eqn.4.9 (Chapter 4) it is clear that the maglev transfer function practically depends on the incremental change of coil-inductance which changes with the operating air-gap of the levitated object. The change of inductance of the magnet-coil with the operating air-gap for the proposed system (described in Chapter 4) is shown in Figure 2 in Chapter 4. The ferromagnetic cylindrical rod contributes to the inductance of the electromagnet coil. As the rod approaches the magnet-coil inductance goes up. As the rod moves farther from the electromagnet, the inductance decreases, reaching a minimal value when it is too far. This minimum value is, of course, the inductance of the magnet-coil.

From Figure 6, it is seen that with the increase of coil-inductance (when the air-gap between the electromagnet and ferromagnetic rod reduces), the system transient response becomes poorer showing more overshoots and undershoots before reaching to the final steady-state point. With the reduction of coil-inductance (when air-gap increases), the system damping is increased. In the frequency domain analysis (Figure 7 & Figure 8), with the increase of inductance, both the GM and the PM is decreased but the GCF is slightly increased showing the fastness of the system.

Figure 6. Effect of change in coil inductance on position responses

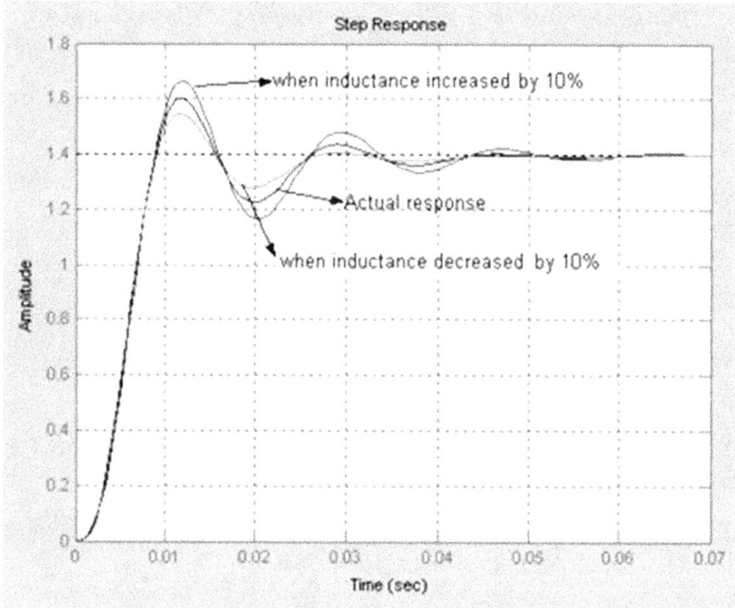

Figure 7. Frequency response at 10 mm air-gap with nominal inductance

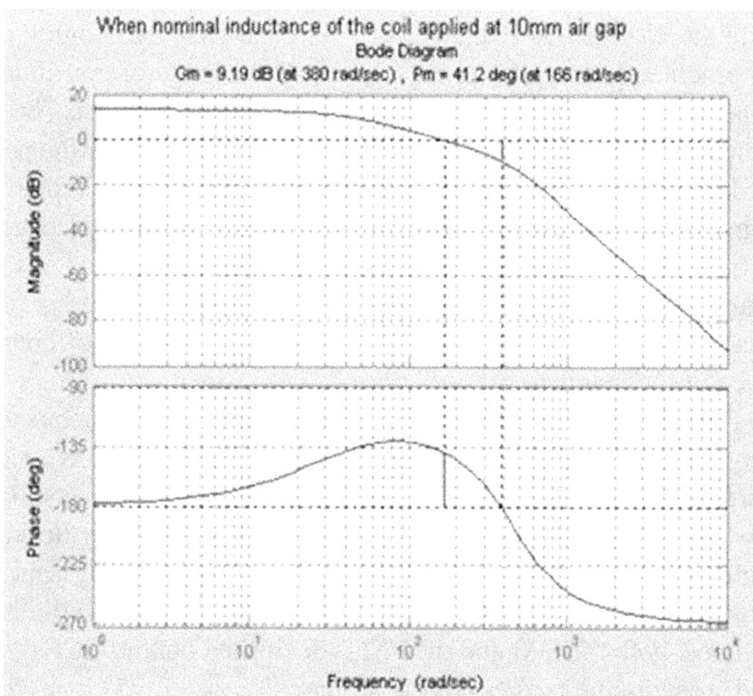

Figure 8. Frequency response at 10 mm air-gap when inductance increased by 20%

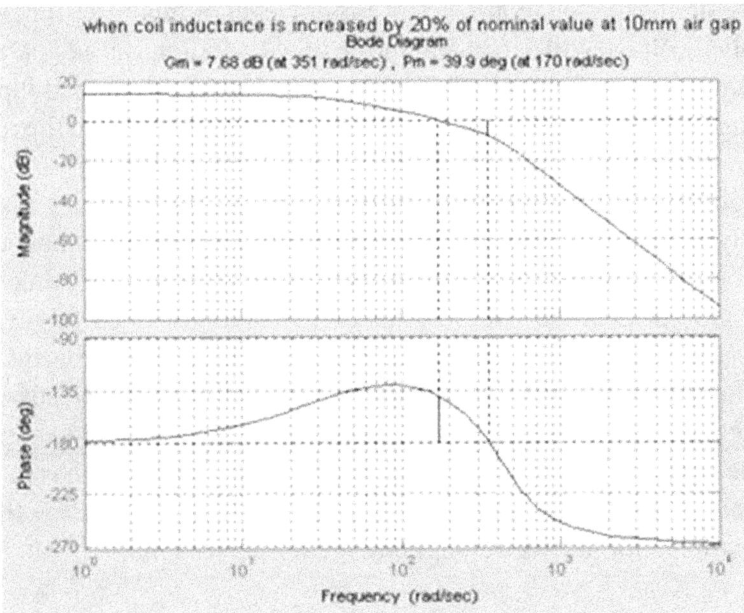

EFFECT OF CHANGE IN INPUT DC LINK VOLTAGE

In the present set-up the magnet-coil is excited by a DC-DC switched mode chopper amplifier. The input DC voltage of the power amplifier is supplied from current limited DC regulated power supply. The input DC link voltage applied to the converter depends on the operating point of the levitated object. By increasing the input voltage, the coil-current and hence the magneto motive force (mmf) can be increased for the same number of turns of the magnet. This method of reducing the effective time constant of the magnet winding is called voltage forcing; the maximum voltage that can be applied is related to the maximum value of $\frac{dI}{dt}$ that can be achieved in a given magnet (slew rate) and the coil inductances (Sinha, 1987).

$$V = L\left[\frac{dI}{dt}\right] \tag{6.6}$$

Where, V is the input DC link voltage appearing across the magnet coil as soon as the switch of the chopper is on and L is the inductance of the magnet.

In the proposed scheme for obtaining stable suspension of the ferromagnetic rod at a desired gap, one requires a minimum level of input DC-link voltage. This applied voltage will depend on the desired levitating gap and parameters of the controller. From Figure 9, it is observed that if the applied input DC-link voltage is higher, the dynamic response becomes better than the response with a lower input DC voltage.

At higher input DC-link voltages, the current loop becomes faster and essentially the tracking capability of the overall system is enhanced. The switched mode chopper gain is also decided by the input DC link voltage and this gain affects the overall gain of the system. Since the system is conditionally stable, the upper value of voltage is restricted by the design value of loop gain. Though use of higher input DC link voltage leads to better control action it also leads to increased EMI problem and higher voltage stress across the switches and diodes of the power circuit. Thus one has to make a compromise between these two conflicting factors while selecting the input DC link voltage for an electromagnetic levitation system.

Figure 9. Effect of input DC link voltage on position response

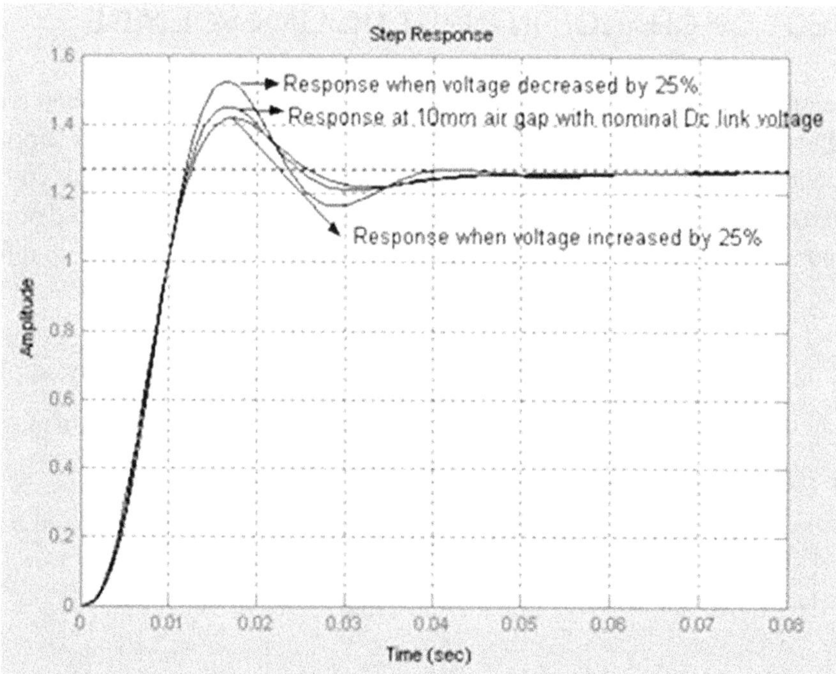

EFFECT OF CHANGE IN CONTROLLER GAIN

The electromagnetic levitated system is inherently unstable and strongly non-linear in nature. To stabilize the system in close loop a Lead controller has been utilized. The transfer-function of the designed controller for 10 mm operating air-gap can be found in Table 4 in Chapter 4.

Without changing the location of pole and zero of the Lead controller, the gain of the controller is varied and the dynamic performance has been observed (Figure 10). It is seen that with the increase of gain the system dynamic response becomes poorer with increasing overshoot and settling time. As expected, increment in gain reduces the steady-state error. From the simulation results, it is obvious that decreasing the gain at a particular operating condition can enhance the system damping sacrificing the system steady state accuracy. Nevertheless, there is a range of gain for which the system performance is acceptable and this range depends on the operating point of the levitated system, parameters of the controller and the nature of step input (magnitude and frequency) applied (Banerjee & Bhaduri, 2008).

Figure 10. Effect of controller gain on position response

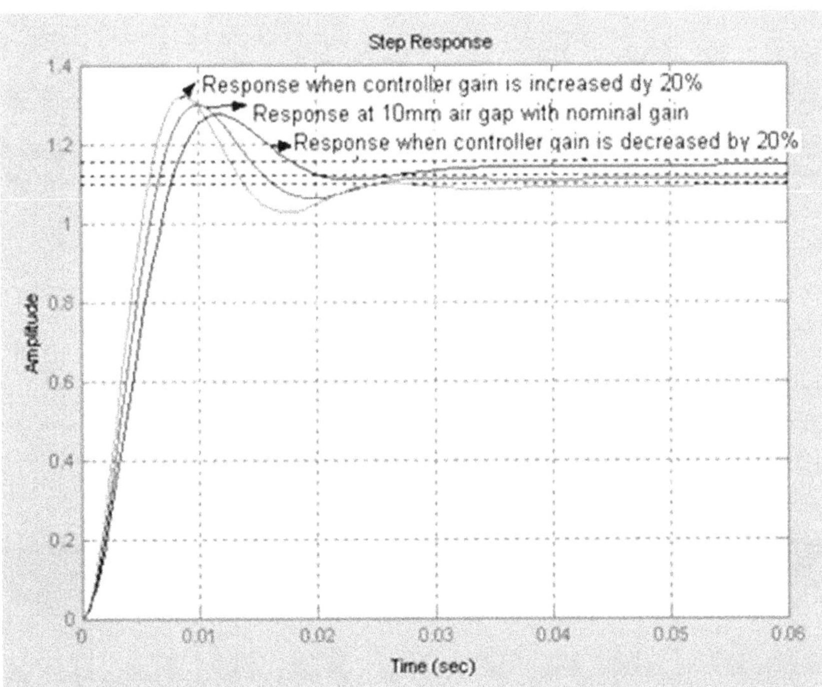

EFFECT OF CHANGE IN POLE LOCATION OF LEAD CONTROLLER

The designed Lead controller for 10 mm air-gap position has zero at -31, and a pole at -369 and the gain of the system is 15.3. Keeping gain and zero constant, the pole location is changed along the negative real axis. When the pole is moved towards the origin, the bandwidth of the controller decreases and the system damping reduces thereby increasing the oscillation in the dynamic response. When the location of the pole is increased keeping the zero and gain constant, the sytem exhibits better damping. But the steady state accuracy is hampered. The designed controller is of course maintaing the both the transient and steady-state performance as seen in Figure 11. It has to be mentioned here the pole of the Lead controller is used to remove the high frequency noise signals which may originate from switched mode power amplifier.

Figure 11. Effect of controller pole on position response

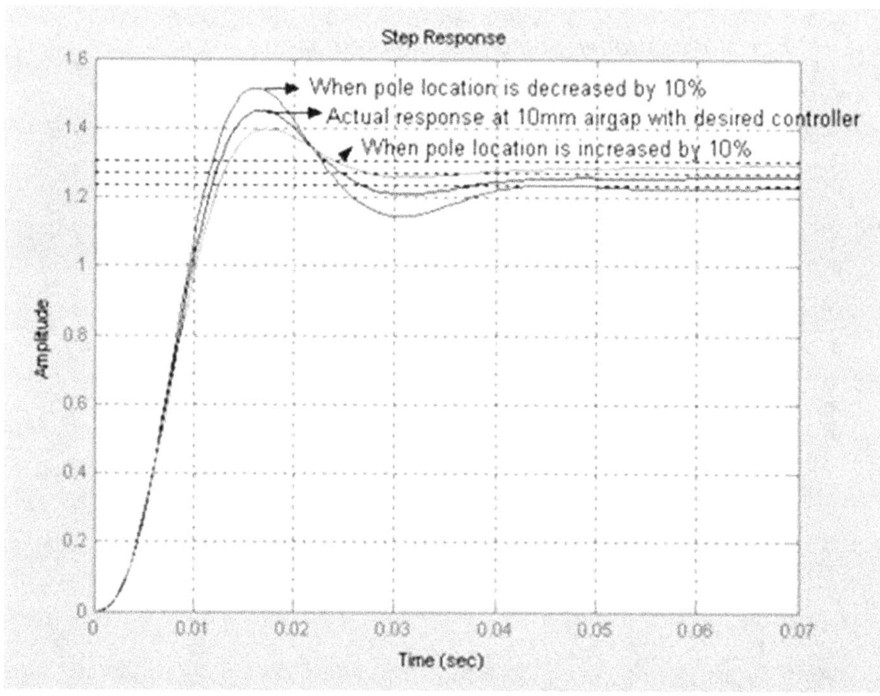

EFFECT OF CHANGE IN ZERO LOCATION OF LEAD CONTROLLER

As described in Chapter 4 (from root locus plot in Figure 10) the closed loop stability of such unstable system may be obtained just adding one zero to the maglev transfer-function which essentially leads to a PD controller. But as stated earlier to remove the high frequency noise signal a pole is also to be added (at least 10 times to the zero location) to the controller transfer-function. So the effect of zero is pre-dominant for maintaining overall close loop stability for such systems.

From Eqn.6.5, the damping ratio will increase with the decrease of zero location on the negative real axis and vice-versa (keeping other parameters constant). So if the zero position is shifted towards left from the designed value then the system will be more oscillatory and vice-versa. This result has been verified with simulation results as shown in Figure 12. With the increase of zero-location the overshoot has been increased, but at the same time system has been little faster and the staedy-state accuracy is slighly improved.

Figure 12. Effect of controller zero on position response

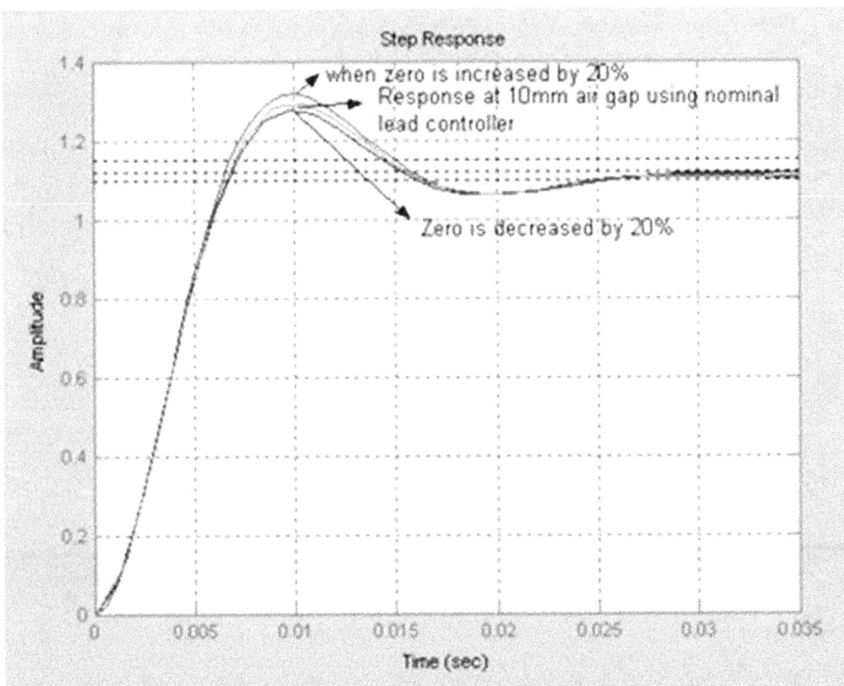

REFERENCES

Banerjee, S., & Bhaduri, R. (2008, December 11–13). The effect of different parameters on the controller performance for a DC attraction type levitation system-a simulation study. In I. I. T. Kanpur (Ed.) *IEEE Conference (INDICON 2008)* (pp. 87–91). 10.1109/INDCON.2008.4768806

Ogata, K. (2000). *Modern control engineering* (3rd ed.). Prentice Hall of India.

Sinha, P. K. (1987). *Electromagnetic suspension, dynamics and control.* Peter Peregrinus Ltd.

Chapter 7
A Double Actuator-Based DC Attraction Type Levitation System for the Suspension of a Cylindrical Rod

ABSTRACT

In the earlier chapters, an attempt has been made for stable suspension of a cylindrical rod under single electromagnet controlling single-axis movement. During experimentation it has been observed that the rod gets tilted to one side that exerts more levitating force due to non-uniformity of the distributed field flux. Moreover, for some specific industrial applications (like induction heating, manufacturing industry, active magnetic bearing, precision instrumentation, mechatronics, etc.), it is required to levitate such cylindrical rods with better pitching control where both ends may be controlled independently. Obviously, for controlling the other degrees of freedom movement of the cylindrical rod, at least two electromagnets are necessary.

INTRODUCTION

In the earlier chapters, an attempt has been made for stable suspension of a cylindrical rod under single electromagnet controlling single-axis movement. During experimentation it has been observed that the rod gets tilted one side that exerts more levitating force due to non-uniformity of the distributed field

DOI: 10.4018/978-1-6684-7388-7.ch007

flux. Moreover, for some specific industrial applications (like induction heating, manufacturing industry, active magnetic bearing, precision instrumentation, mechatronics etc.) it is required to levitate such cylindrical rod with better pitching control where both end may be controlled independently. Obviously, for controlling the other degrees of freedom movement of the cylindrical rod, at least two electromagnets are necessary.

This work describes design, implementation and testing of a DC attraction type levitation system where a ferromagnetic cylindrical rod of around 0.122 kg mass is made to remain suspended at the desired operating gap under two I-core electromagnets (Figure 1). The two electromagnets are controlled independently through two identical controllers and the stable levitation of the rod is achieved through single input and single output (SISO) control of each air-gap corner. The emphasis of this work is to design and development of controller unit for two actuators. The single switch based power circuit simplifies the overall hardware and it can be extended to any number of magnet-coil. A cascade compensation control scheme utilizing inner current loop and outer position loop has been designed and implemented for stabilization of such highly unstable and strongly non-linear system. The prototype has been successfully tested and stable levitation was demonstrated at the desired operating gap.

DESCRIPTION OF EXPERIMENTAL SET-UP

The block diagram of individual unit for the proposed DCALS is shown in Figure 1. In each case the current of the electromagnet is controlled through the DC to DC switch mode chopper circuit utilizing an outer position control loop and an inner current feedback control loop. The parameters of the maglev systems are given in Table 7.2. The photograph of the experimental setup is shown in Figure 2.

When the two electromagnets are simultaneously excited, a net attractive force is generated between the magnet pole-faces and the ferromagnetic rod, as a result of which the magnets try to pull up the complete ferromagnetic rod. The dedicated independent controller used for each magnet tries to control the air-gap between that magnet pole-face and the cylindrical rod by maintaining the required current in the corresponding magnet-coil. With each magnet cum controller unit working satisfactorily, each side of the cylindrical rod gets the desired vertical lift and in the process the whole cylindrical rod is levitated.

The magnet current is controlled by a single switch based DC to DC chopper circuit (Figure 24 in Chapter 8). Since the electromagnetic levitation system is inherently unstable, the selection and design of the controller is important so that the overall closed loop system becomes stable and gives satisfactory performance. A cascade compensator is used with the position control loop for maintaining overall closed loop stability.

Two linear inductive type position sensors have used to measure two air-gaps (gap between the magnet pole-face and cylindrical rod). Output of the position loop (current reference signal) is compared with the actual coil-current signal sensed by an LEM make (LA-55P) Hall-effect current sensor. For better dynamic response and steady state accuracy the current error is processed through a PI controller and its output is used to control the chopper output voltage through PWM control logic. The duty ratio of the MOSFET switches varies as the object moves up and down within the electromagnetic field. When the rod moves upwards (beyond reference gap) the duty ratio of the MOSFET gate pulse is decreased, consequently the magnet current decreases and causes it to go down and vice-versa.

The stable suspension of the prototype is achieved through the control of air-gap between the magnet and the cylindrical rod by controlling the current flowing through the magnet-coil. The two magnets are controlled independently through two identical controllers and stable levitation of the rod is achieved through SISO control of each air-gap. The steady-state position of each end of the rod may be controlled through each reference gap command and has been demonstrated experimentally.

Figure 1. Schematic diagram of individual unit for the proposed DCALS

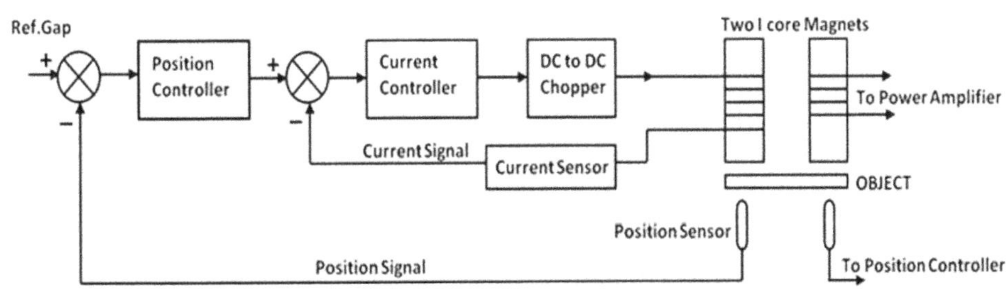

Figure 2. Photograph of the experimental setup

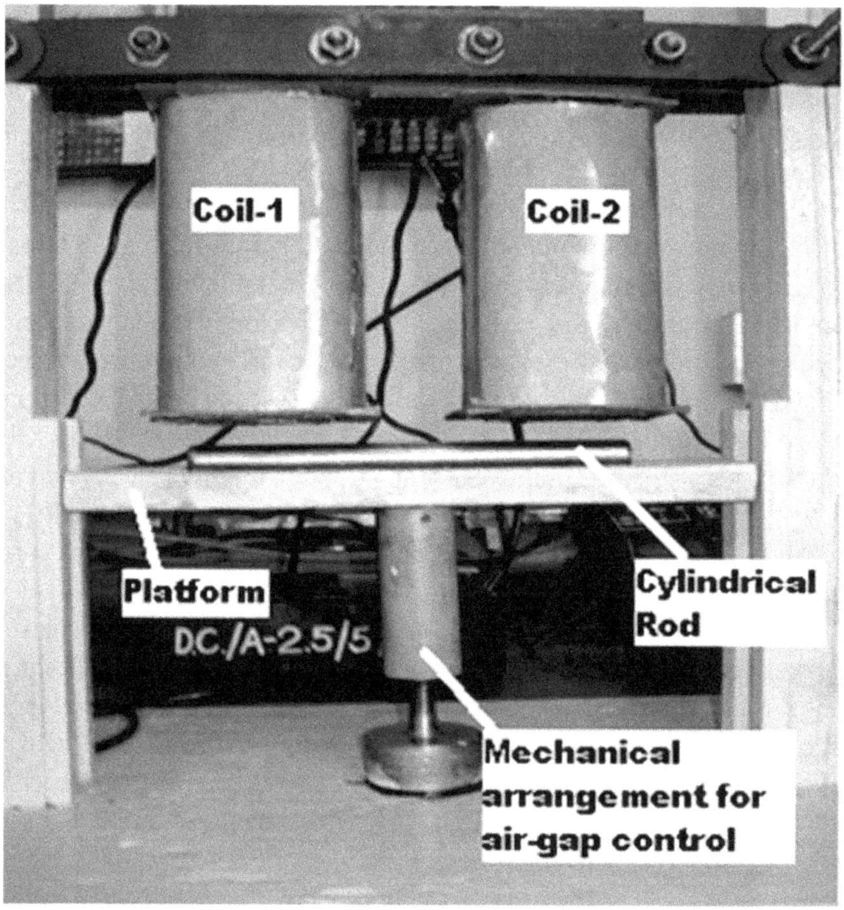

FEM SIMULATION

Finite element method (FEM) based ANSYS (version 12.1) analysis has been done to find the flux pattern, working flux density, field intensity etc. for the two-magnet based levitation system

ANSYS is engineering simulation software that has many finite-element analysis capabilities, ranging from a simple, linear, static analysis to a complex, nonlinear, transient dynamic analysis (ANSYS, Inc. 2009).

The procedure for doing the FEM analysis utilizing ANSYS software of any project consists of five main steps:

1. Creation of the physics environment.
2. Building and meshing the model and assigning physics attributes to each region within the model.
3. Applying boundary conditions and loads (excitation).
4. Obtaining the solution.
5. Reviewing the results.

The flow-chart for the ANSYS simulation procedure for the proposed system is shown in Figure 3.

Figure 3. Flow-chart for the ANSYS simulation procedure

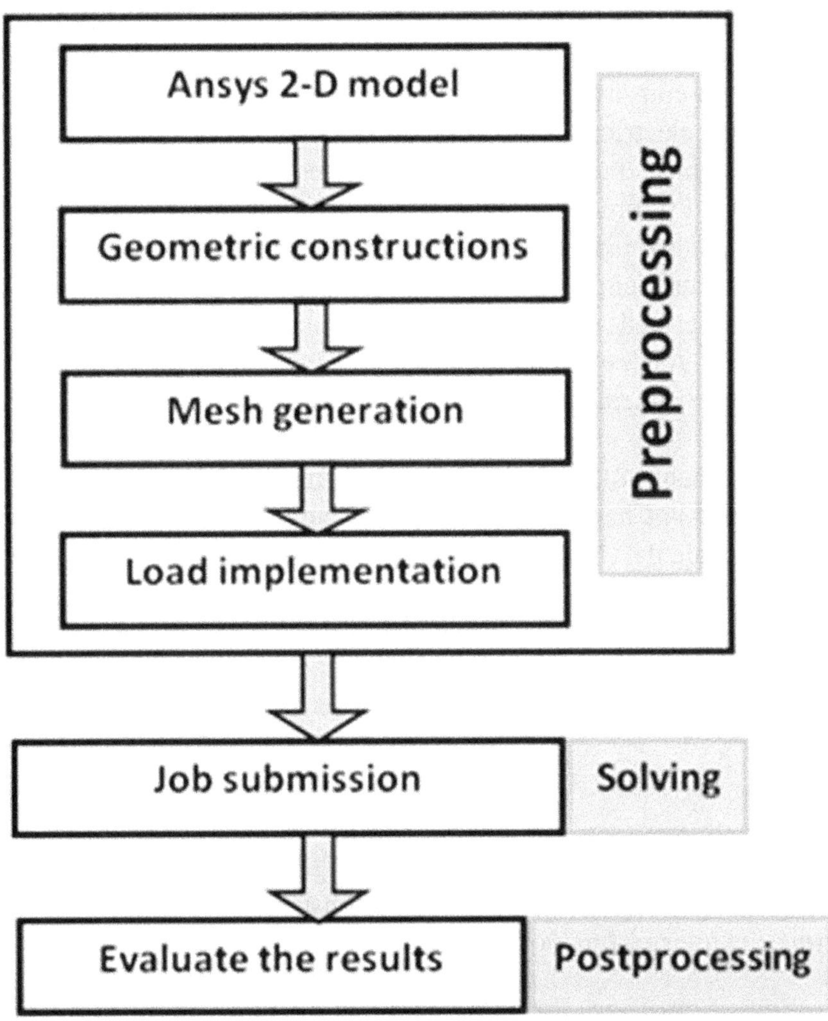

Figure 4 shows the ANSYS model of the prototype for two-dimensional analysis. The electromagnets have been excited by the rated current (0.52A) at the desired operating gap (10 mm) under static condition. As per the set-up, the system may be considered as two magnetically decoupled (in ideal case) units. But here, the problem has been treated as a two-dimensional problem in the centre plane on which the axes of the two adjacent electromagnets lie. The flux pattern and field intensity of the system at 10mm operating air-gap using FEM software ANSYS is shown in Figure 5 and Figure 6. The flux pattern and field intensity etc. is symmetrical for the both sections of the system.

ESTIMATION OF SYSTEM PARAMETERS AND DETERMINATION OF PLANT TRANSFER-FUNCTION

The two magnet-coils used in this work are identical each having almost same mechanical and electrical parameters. The mass of the cylindrical rod is 122 gm and it is so placed under the two I-core electromagnets (Figure 2) such that the upward lift force to be generated by each magnet is equivalent to 61 gm (assuming uniform distribution of the load) and this leads to estimation of magnet parameters being derived by the technique used in the single magnet levitation as described in Chapter 4. The parameters for the present set-up are given in Table 2. As described in Chapter 4 the inductance and current profile for the two independent systems have been determined experimentally and since both the systems are almost identical here only one set of characteristics (Figure 7 and Figure 8) for coil-1 have been presented.

As described in Chapter 2, for controlled current source excitation input to the magnet coil, the transfer-function of the maglev plant can be expressed by Eqn.7 1.

$$G_p(s) = \frac{\Delta Z(s)}{\Delta I(s)} = -\frac{\left(\frac{K_a}{m}\right)}{\left(s^2 - \frac{K_X}{m}\right)} \tag{7.1}$$

After finding the values of the two force constants (K_a and K_X) for a known mass (m), the transfer function of the plant can be determined.

Figure 4. ANSYS model of the prototype for two-dimensional analysis

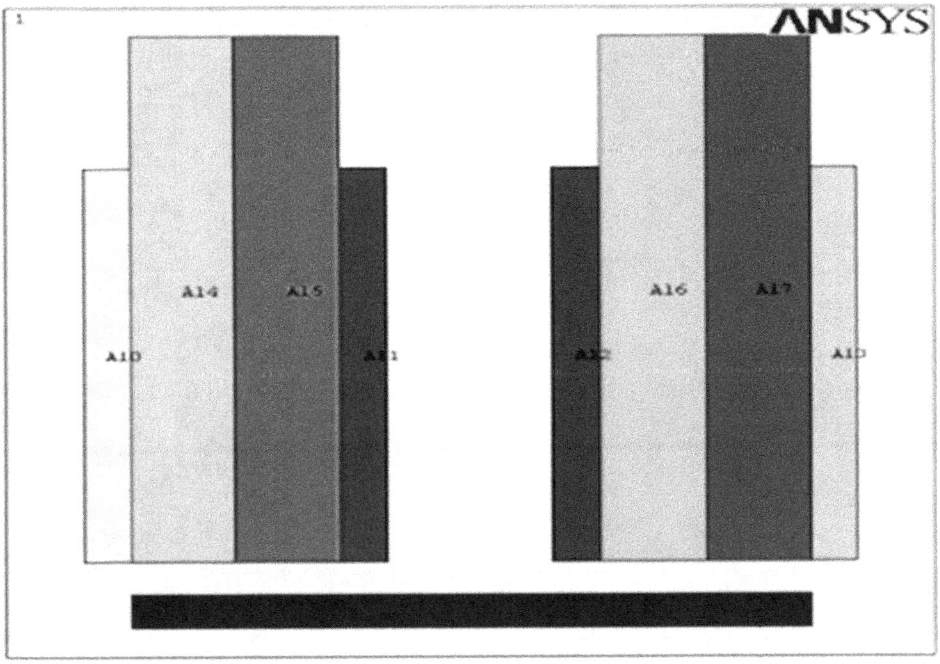

Figure 5. Flux pattern for the system at 10 mm air-gap

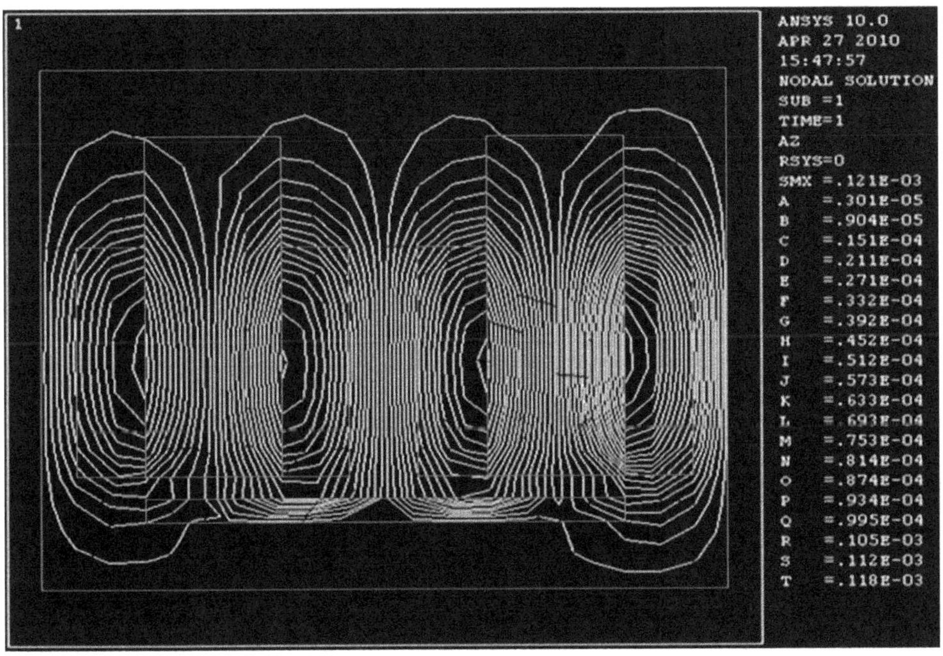

Figure 6. Field intensity plot of the system at 10 mm air-gap

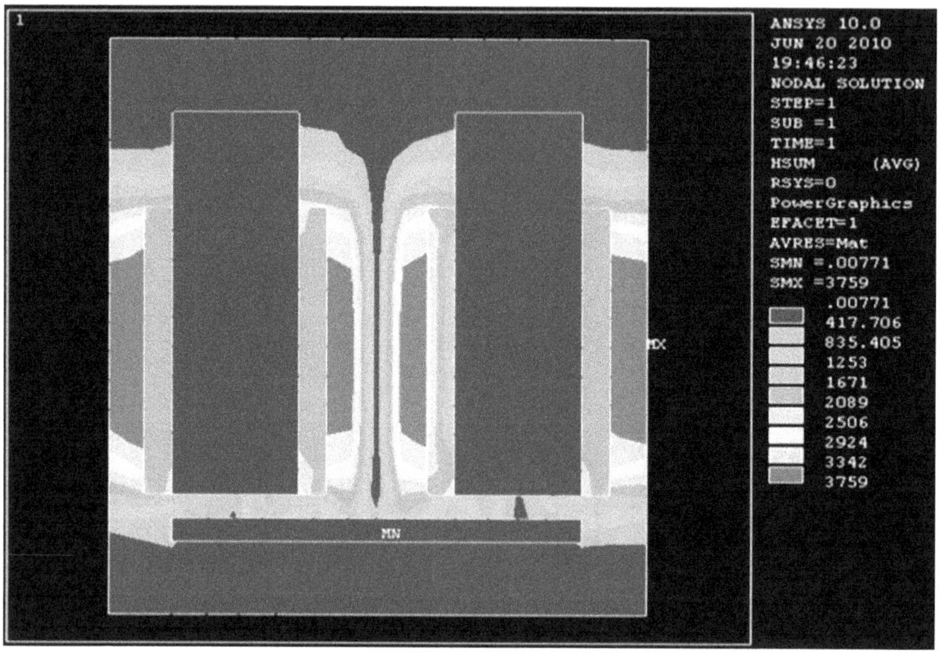

Figure 7. Inductance profile of the Coil-1

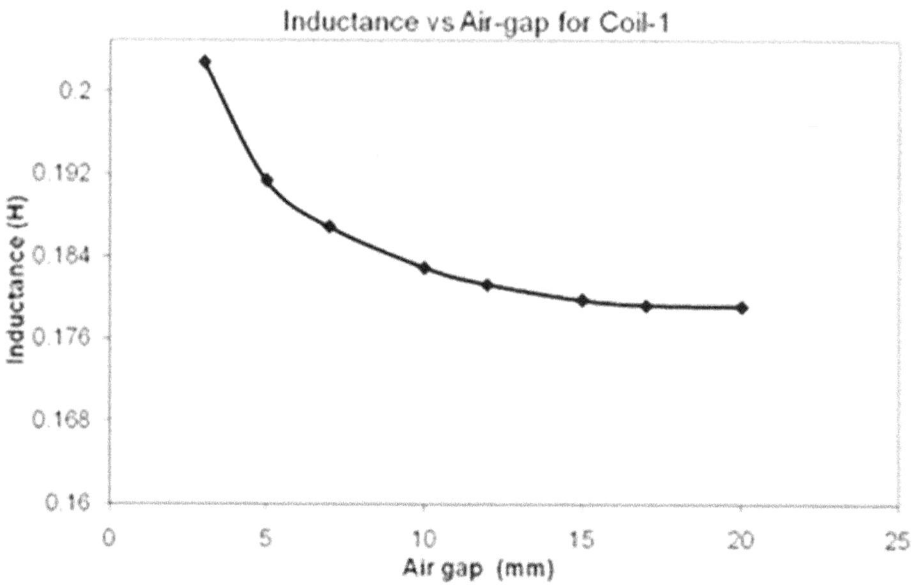

Figure 8. Pick up current vs. air gap for Coil-1

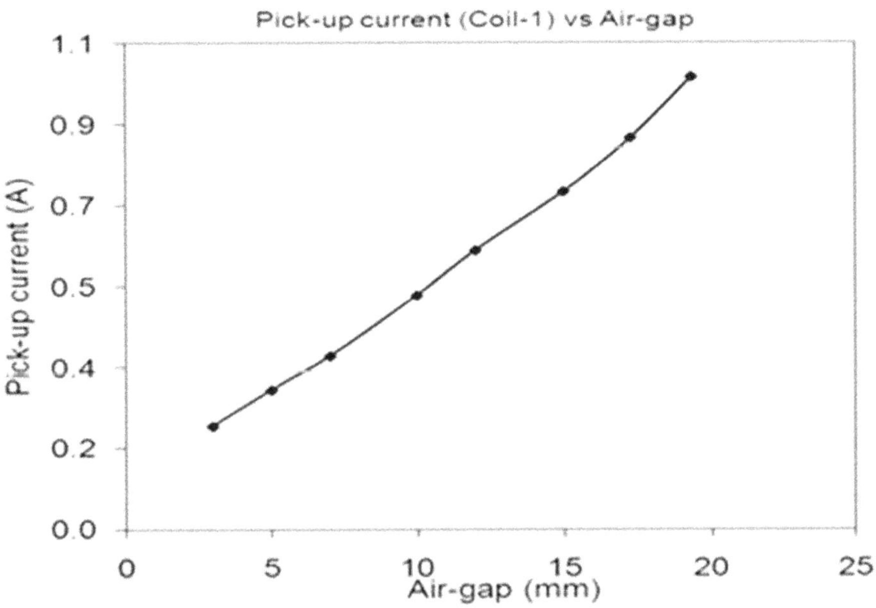

Sample Calculation (Operating Air-Gap of 10mm for Coil-1)

The incremental inductance (L_0) = Total inductance (at air-gap 10 mm) – Coil self-inductance

= (0.1829−0.16648) H ≅ 0.0164 H

The force constants are determined as follows:

$$C = \frac{L_0 x_0}{2} = \frac{0.0164 \times 10 \times 10^{-3}}{2} = 0.000082 \tag{7.2}$$

$$K_a = 2C\left(\frac{i_0}{x_0^2}\right) = 2 \times 0.000082 \times \left(\frac{0.52}{\left(10 \times 10^{-3}\right)^2}\right) \cong 0.853\ N / A \tag{7.3}$$

$$K_X = 2C\left(\frac{i_0^2}{x_0^3}\right) = K_a\left(\frac{i_0}{x_0}\right) = 0.853 \times \left(\frac{0.52}{10\times 10^{-3}}\right) \cong 44.36\ N/m \tag{7.4}$$

So the transfer function of the plant (at operating air-gap 10 mm) can be found from Eqn.7.1,

$$G_p(s) = \frac{\Delta X(s)}{\Delta I(s)} = \frac{\left(\frac{K_a}{m}\right)}{\left(s^2 - \frac{K_X}{m}\right)} = \frac{\left(\frac{0.853}{61\times 10^{-3}}\right)}{\left(s^2 - \frac{44.36}{61\times 10^{-3}}\right)} = \frac{13.98}{\left(s^2 - 727.213\right)} = \frac{13.98}{(s+26.97)(s-26.97)} \tag{7.5}$$

Table 1. Transfer functions of maglev plant (coil-1) at different air-gaps

Air Gap (mm)	Current (Amp)	Total Inductance (Henry)	Incremental Inductance (Henry)	K_a (N/A)	K_X (N/m)	Transfer-Function of Plant
3	0.23	0.2029	0.0364	2.79	213.922	$\frac{45.74}{(s+59.22)(s-59.22)}$
5	0.31	0.1914	0.0249	1.544	95.744	$\frac{25.32}{(s+39.62)(s-39.62)}$
7	0.385	0.1901	0.0236	1.3	71.511	$\frac{21.32}{(s+34.24)(s-34.24)}$
10	0.52	0.1829	0.0164	0.853	44.373	$\frac{13.98}{(s+26.97)(s-26.97)}$
12	0.62	0.1822	0.0157	0.812	41.977	$\frac{13.32}{(s+26.23)(s-26.23)}$
15	0.795	0.1807	0.0142	0.755	39.989	$\frac{12.37}{(s+25.60)(s-25.60)}$
17	0.87	0.1793	0.0128	0.656	33.576	$\frac{10.76}{(s+23.46)(s-23.46)}$
20	0.963	0.1792	0.0127	0.612	29.490	$\frac{10.04}{(s+21.98)(s-21.98)}$

The transfer-functions of maglev plant (coil-1) at different air-gaps have found experimentally and are given in Table 1.

Table 2. System data

Parameters	Values
Current sensor gain	1
Mass of the object	122 gm
No. of electromagnets	2
Effective mass of object per magnet	61 gm
Length of the rod	14.5 cm
Diameter of the rod	10 mm
Resistance of the coil-1	4.1 Ohm
Resistance of the coil-2	3.9 Ohm
Self-Inductance of the coil-1	0.16648 H
Self-Inductance of the coil-2	0.16597 H
Position sensor gain	1000 V/m

DESIGN OF CONTROLLERS

In this work two similar controllers have been used for controlling two independent actuators and the basic schematic diagram of the overall closed loop system for one unit is shown in Figure 9. Essentially independent system utilizes two-control loops, an inner current loop and an outer position loop similar to a single actuator based system as described in Chapter 4 and design of controllers have also been done using similar approach.

Figure 9. Basic block diagram of overall closed loop system for DCALS

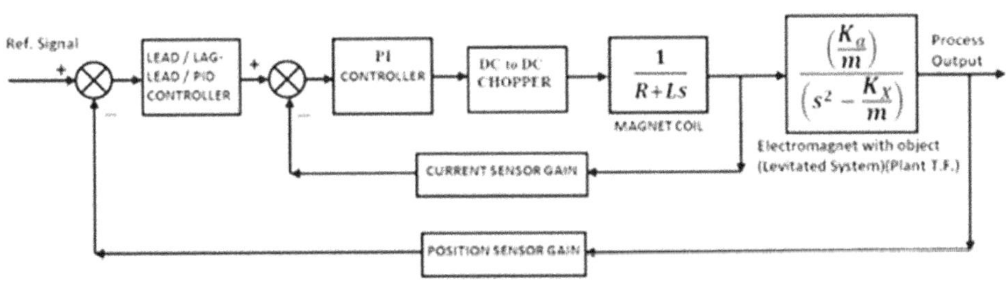

Current Control Loop

To design a suitable current controller for providing better dynamic response it is necessary to model the current control loop accurately. The basic block diagram of the current control loop (Figure 10) consists of different blocks such as current controller, chopper amplifier, magnet inertia and current sensor in the feedback path.

Current Controller

The current error signal is fed into a controller, which could be a proportional (P), proportional plus integral (PI), or proportional, integral, and differential (PID). The most commonly used controller among them is the PI controller that gives better dynamic performance and ensures zero steady-state error. In this work a proportional plus integral (PI) is used to regulate the current. The transfer function of the current controller is given by

$$G_{cc}(s) = K_p + \frac{K_i}{s} \tag{7.6}$$

Where, K_P and K_i are the proportional and integral gains of the controller.

Power Amplifier

As described in Chapter 4 the chopper amplifier can be modelled by a gain and a time constant which can be represented by the following transfer function.

$$G_{ch}(s) = \frac{K_{ch}}{(1 + sT_c)} \tag{7.7}$$

Where K_{ch} is the gain and T_c is the time constant of the chopper circuit.

Here, the PI controller output is used to control chopper duty ratio such that the output from PI controller and mean DC voltage output by chopper are proportional. The chopper output voltage thus varies linearly with the PI controller output. With an input supply of around 100 volts, and control voltage of 5 volts the chopper gain (ratio between output voltage and control voltage coming from PI controller) will be equal to 20. The time lag in the converter is equal to the average carrier switching-cycle time, i.e., half the period, and is expressed in terms of the PWM switching frequency as $T_c = \frac{1}{2f_c}$; In this case the switching frequency of the converter is taken to be 27 kHz. So the delay angle is about 0.0000185 ms. so the transfer function of the chopper amplifier becomes:

$$G_{ch}(s) = \frac{20}{(1 + .0000185s)};$$

After neglecting the very small time constant of the chopper amplifier, the transfer function of the same becomes a simple gain of 20.

Magnet Coil

Taking the Laplace transform of Eqn.4.16 (Chapter 4) and after rearranging the transfer function of the magnet-coil appears as:

$$\frac{I(s)}{V(s)} = \frac{1}{(R + sL\ )} \tag{7.8}$$

Experimentally, the transfer function of the coil at the nominal operating point (10mm) gap is found to be

$$\frac{I(s)}{V(s)} = \frac{1}{(4.1 + 0.1829s)} \tag{7.9}$$

Current Sensor

In this work an LEM make current sensor (LA-55P) was used to measure the actual coil current. The bandwidth of the sensor is very high (1 MHz) and there is perfect (for our purpose) linearity between the actual current and the sensed current signal. A voltage signal proportional to the actual current appears across an output resistor connected between the sensor output and the ground. In this measurement the current sensor was scaled to give an output of 1.0 volts per ampere of coil current. So the feedback path gain (K_c) for the current loop is equal to 1.0.

Design of Current Controller

The block diagram of the current control loop showing the different subsystems in transfer function form is shown in Figure 10.

Figure 10. Block diagram of current loop

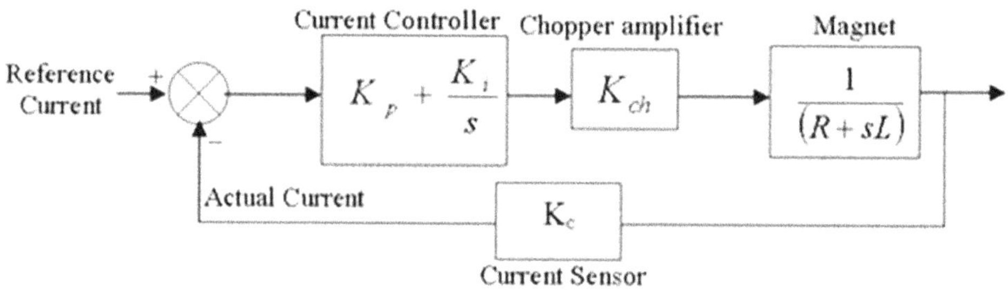

Thc loop gain is written as

$$G(s)H(s) = K_{ch}K_c(K_p + \frac{K_i}{s}) \times \frac{1}{(R+sL)} = \frac{K_{ch}K_cK_i}{R} \frac{(1+s\frac{K_p}{K_i})}{s} \times \frac{1}{(1+s\frac{L}{R})} \tag{7.10}$$

For the given parameters of the system the gains of the PI controller can be determined both in the frequency and time domain.

Frequency Domain Approach

After converting the above Eqn.7.10 in the frequency domain, from the phase angle criterion one can write:

$$-90^o + \tan^{-1}\left(\frac{\omega K_p}{K_i}\right) - \tan^{-1}\left(\frac{\omega L}{R}\right) + 180^o = \Phi \tag{7.11}$$

Where Φ, the phase margin (PM) is assumed to be around 60 deg. for good margin of stability. The high value of PM is chosen, because in an experimental implementation additional time lag or saturation decreases the phase margin anyway. By choosing a large PM, stability can be guaranteed for a wide range of operation. The bandwidth of the current feedback loop is to be set higher than the gain crossover frequency of the position control loop.

Sample Calculation:-

Gap = 10 mm for Coil-1:

L = 0.1829 H, R = 4.1Ω.

$$G(s)II(s) - K_{ch}K_c(K_p + \frac{K_i}{s}) \times \frac{1}{(R + sL)} \tag{7.12}$$

$$G(s)H(s) = K_{ch}K_c\frac{\left(K_p s + K_i\right)}{s} \times \frac{1}{(R + sL)}$$

$$or, G(j\omega)H(j\omega) = \frac{K_{ch}K_c\sqrt{\omega^2 K_p^2 + K_i^2}}{\omega\sqrt{R^2 + \omega^2 L^2}}$$

$$-90^o + \tan^{-1}\left(\frac{\omega K_p}{K_i}\right) - \tan^{-1}\left(\frac{\omega L}{R}\right) + 180^o = \Phi$$

We know that the P.M should be 60^0.

So, P.M = $180^0 + \Phi$

$$\text{Or, } -90^o + \tan^{-1}\left(\frac{\omega K_p}{K_i}\right) - \tan^{-1}\left(\frac{\omega L}{R}\right) + 180^o = 60^o$$

$$\text{Or, } \frac{K_p}{K_i} = 0.0025427 \tag{7.13}$$

Now, $|G(s)H(s)|=1$ where, gain cross over freq. GCF in Hz.=100 Hz.

Or, $\dfrac{K_{ch}K_c\sqrt{\omega^2K_p^2+K_i^2}}{\omega\sqrt{R^2+\omega^2L^2}}=1$

Or, $$\frac{20\sqrt{394384K_p^2+K_i^2}}{628\sqrt{16.81+394384\times(0.03345)}}=1 \quad (7.14)$$

$K_p = 8.1228$

$K_i = 3194.5$

Here, choosing the gain crossover frequency of the current loop to be 100 Hz (which is taken at least ten times of the allowable vibration frequency of suspension system that is around 10 Hz) one can find the proportional gain (K_p) and integral gain (K_i) of the current controller from Eqn.7.13 and Eqn.7.14. In this design after substituting the values of the different parameters the value of K_p(=8.1228) and K_i(=3194.5) is determined.

From Eqn.7.12 the loop gain becomes

$$G(s)H(s)=\frac{15582.92(1+0.0025427\text{s})}{s(1+0.0446s)} \quad (7.15)$$

The Bode plot (Figure 11) of the current loop verifies the success of the design. Here the achieved phase margin is 60 degree at a gain crossover frequency of 100 Hz. The system damping ratio can be calculated approximately by the relation ζ=0.01Φ= 0.6 which is also verified in the time domain approach.

Figure 11. Frequency response of the current loop

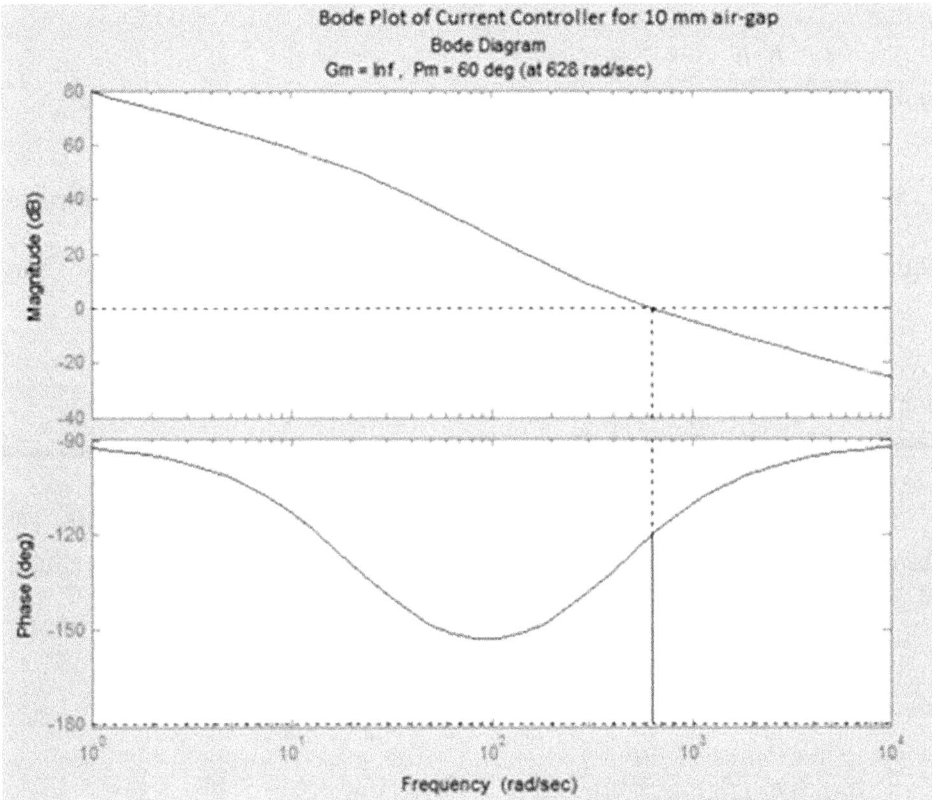

Time Domain Approach

The characteristic equation of the current loop is

$$1+G(s)H(s)=1+\frac{K_{ch}K_{c}(sK_{p}+K_{i})}{s(R+sL)}=0 \tag{7.16}$$

or, $s(R+sL)+(sK_p+K_i)K_{ch}K_c=0$;

$$\text{or,}\ L\left[s^{2}+s\left\{\frac{R}{L}+\frac{K_{p}K_{ch}K_{c}}{L}\right\}+\frac{K_{i}K_{ch}K_{c}}{L}\right]=0$$

Since L is not equal to zero,

$$s^2 + s\left\{\frac{R}{L} + \frac{K_p K_{ch} K_c}{L}\right\} + \frac{K_i K_{ch} K_c}{L} = 0 \tag{7.17}$$

By comparing with the prototype second order time domain characteristic equation, the natural frequency becomes, $\omega_n = \sqrt{\frac{K_i K_{ch} K_c}{L}}$ rad/sec. and damping ratio $\zeta = \frac{1}{2}\frac{\left\{\frac{R}{L} + \frac{K_p K_{ch} K_c}{L}\right\}}{\sqrt{\frac{K_i K_{ch} K_c}{L}}}$

The unit step response of the current control loop is given in Figure 12. As expected the plot shows a good dynamic response having an overshoot of 20% and zero steady state error as expected.

Figure 12. Unity step response for 10 mm of current loop

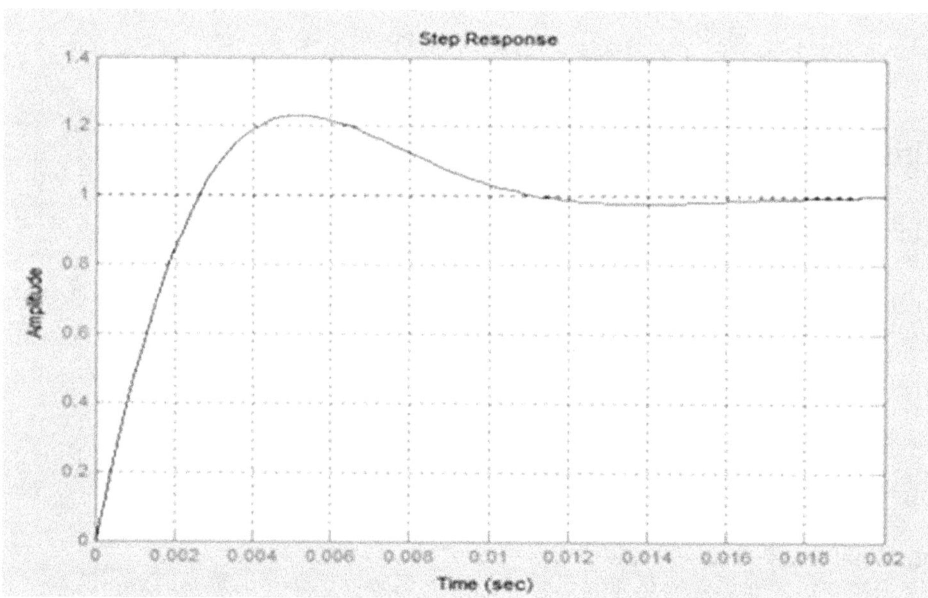

Design of Position Controller

For deigning the slower position loop controller, inner current loop dynamics (which is at least 10 times faster than position loop) may be neglected or approximated to a unity gain block. The basic block diagram of the simplified position control loop is shown in Figure 13.

Figure 13. Block diagram of position control loop

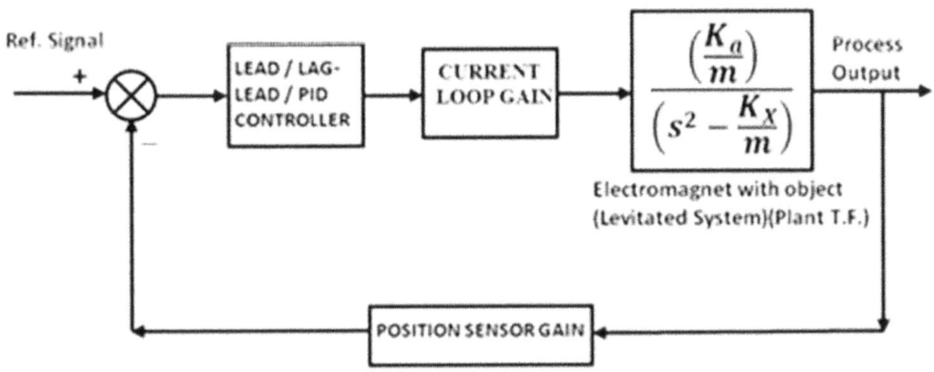

Transfer Functions of the Different Blocks in the Position Control Loop

To design the position controller it is necessary to determine the model of the different subsystems of the position control loop. The transfer function of the levitated system (plant) is given in the following sub-section.

Levitated System (Plant)

The dynamics of the levitated object is considered to be the plant transfer function of the system. Since the controlled current source is taken as the input of the actuator coil, the transfer function of the plant is given as

$$G_p(s) = \frac{\Delta X(s)}{\Delta I(s)} = -\frac{\left(\frac{K_a}{m}\right)}{\left(s^2 - \frac{K_X}{m}\right)} \tag{7.18}$$

For the nominal operating air gap of 10 mm for Coil-1, with m = 61 gm, $K_a = 1.544$(N/A) and $K_X = 95.74$(N/m). After substituting all these values, the transfer function of the plant at 10 mm air gap becomes

$$G_p(s) = \frac{\Delta X(s)}{\Delta I(s)} = \frac{13.98}{\left(s^2 - 727.213\right)} = \frac{13.98}{(s + 26.97)(s - 26.97)} \tag{7.19}$$

REFERENCES

ANSYS, Inc. (2009). *Licensing guide Release 12.1* (002829). Author.

Chapter 8
Implementation of the Double Actuator-Based DCALS

ABSTRACT

Two identical controllers have been utilised for the two electromagnets independently, and the cylindrical rod has been levitated by employing independent SISO control of each gap. The whole unit was considered to be composed of two decoupled identical subsystems. The use of decoupling technique simplified the overall analysis and design of the controller units. The cascade compensators for the two units have been designed by using classical synthesis method. The prototype has been successfully tested, and its stable levitation has been demonstrated at the desired gap position. The two independent SISO controllers designed for controlling and maintaining a fixed air-gap at the two sides of the rod are found to work successfully in unison.

SIMULATED RESULTS

In Chapter 4 design of different classical controllers for single actuator based maglev systems have been discussed in detail. Similar procedure is followed for designing linear controllers for the present two actuator based DCALS and is not being discussed the same in this chapter. Only the simulated results have been presented. Figure 1 and Table 1 shows the performance of designed Lead controllers for different air-gap positions of proposed DCALS. It is seen that with the designed Lead controllers for different operating air-gaps the closed

DOI: 10.4018/978-1-6684-7388-7.ch008

loop system becomes stable and satisfactory transient performances have been achieved but steady-state performance is poor since the Lead controller cannot eliminate the steady-state error. The DC gain of compensators keep on increasing since the plant inherent gain is decreasing with the air-gap. From Table 1 it is clear that for the same percent peak overshoot and steady-state error, the system response becomes sluggish with increased air-gap.

Figure 1. Overall closed loop frequency response with Lead controller at 10mm gap

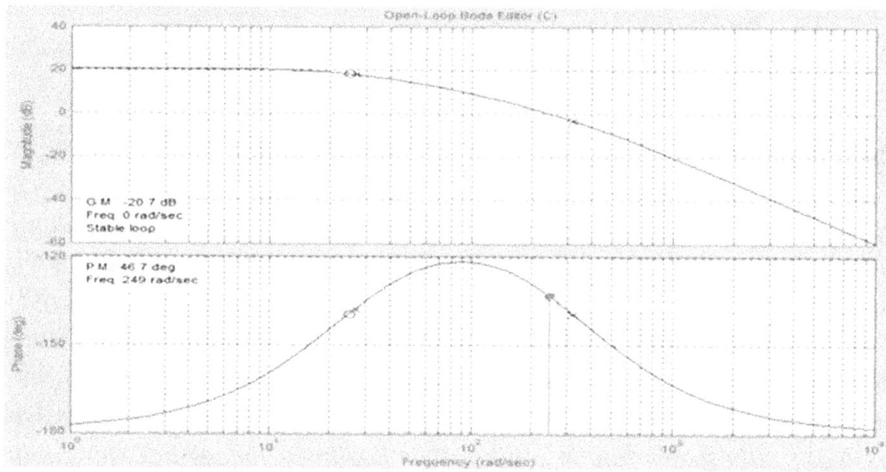

Figure 2. Combined responses with Lead controllers at different operating positions

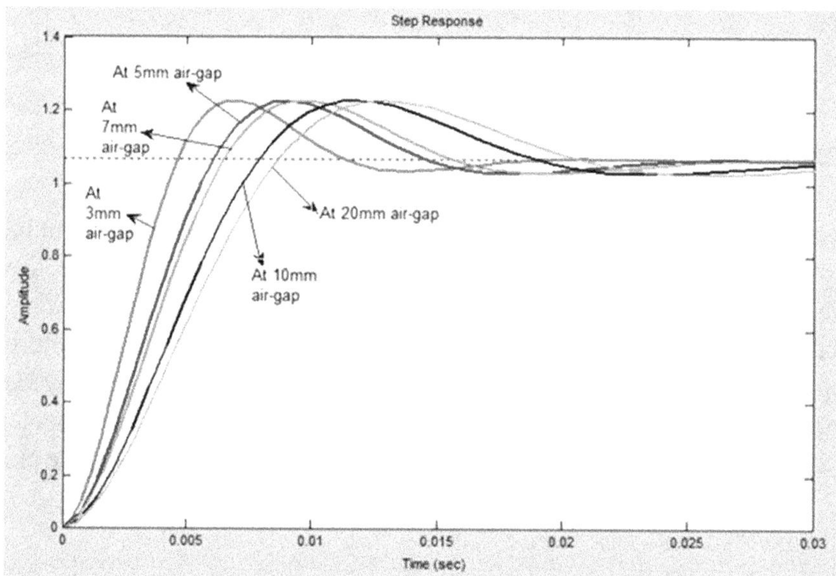

Table 1. Performances of Lead controller at different operating positions

Gap (mm)	T.F. of the Plant	T.F. of LEAD Comp.	(%) Overshoot	Rise Time (sec)	Settling Time (sec)	S.S. Error	Margin of Stability (Kmin to Kmax)	Stable Closed Loop Poles
3	$\frac{45.74}{(s+59.22)(s-59.22)}$	$\left(6.7\frac{s+57}{s+572}\right)$	16	0.00314	0.0161	0.13	0.77 to inf.	-56.5,-258±455i
5	$\frac{25.32}{(s+39.62)(s-39.62)}$	$\left(6.9\frac{s+37}{s+424}\right)$	16	0.00411	0.0216	0.11	0.72 to inf.	-36.5,-194±349i
7	$\frac{21.32}{(s+34.24)(s-34.24)}$	$\left(7.1\frac{s+32}{s+395}\right)$	16	0.00441	0.023	0.11	0.68 to inf.	-31.6,-182±325i
10	$\frac{13.98}{(s+26.97)(s-26.97)}$	$\left(7.3\frac{s+25}{s+323}\right)$	16	0.00535	0.0281	0.1	0.673 to inf.	-24.6,-149±268i
12	$\frac{13.32}{(s+26.23)(s-26.23)}$	$\left(7.6\frac{s+24}{s+320}\right)$	16	0.00537	0.0285	0.1	0.689 to inf.	-23.6,-148±268i
15	$\frac{12.37}{(s+25.60)(s-25.60)}$	$\left(7.8\frac{s+23}{s+309}\right)$	16.1	0.00548	0.0295	0.1	0.72 to inf.	-22.6,-143±262i
17	$\frac{10.76}{(s+23.46)(s-23.46)}$	$\left(8.1\frac{s+21}{s+293}\right)$	16.1	0.00577	0.031	0.1	0.714 to inf.	-20.6,-136±250i
20	$\frac{10.04}{(s+21.98)(s-21.98)}$	$\left(8.3\frac{s+20}{s+289}\right)$	16.1	0.00589	0.0313	0.09	0.696 to inf.	-19.7, -135±244i

The root-locus plot (Figure 3) of overall closed loop system with Lead-Lag compensation at 10 mm operating air-gap shows the system is conditionally stable. The frequency response plot (Figure 4) shows improved performance than that was obtained using Lead compensation. From Table 2, and Figure 5, it is clear that introducing Lead-Lag network satisfactory transient and steady-state performance is achieved irrespective of any operating air-gap. With the introduction of Lag controller, the order of the overall system increases and hence the range of gain for stable operation of the overall system reduces than with the Lead controller. But this controller reduces steady-state error drastically. The advantage of this scheme is that here both pole and zero can be suitably placed to get the desired performance which is not being possible in case of PID controller.

As stated earlier, the DCALS is inherently unstable system; the application of PID controller becomes critical due to having a pole at origin. The system can be stabilized for a smaller value of gain (Table 3) and the margin of stability is also less. But the steady-state error can be completely removed with PID control. From Table 3 it is seen that, DC gain of the controller becomes less than that of Lead-Lag compensator (Table 2). Figure 6 shows the responses of the levitated system with PID controller at different air-gap positions.

Figure 3. Root locus plot for 10 mm plant with Lead-Lag compensation

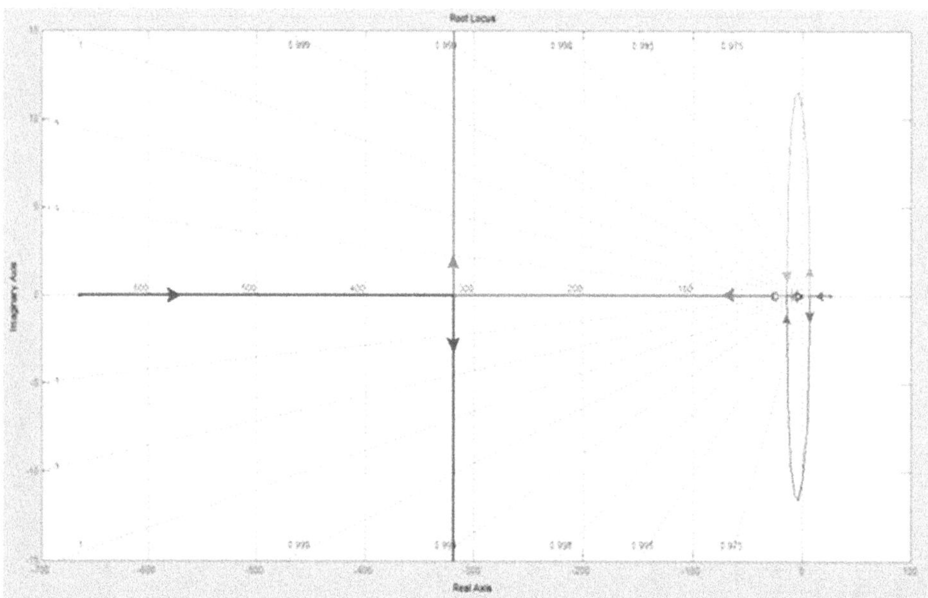

Figure 4. Overall closed loop frequency response with lead-lag controller at 10mm gap

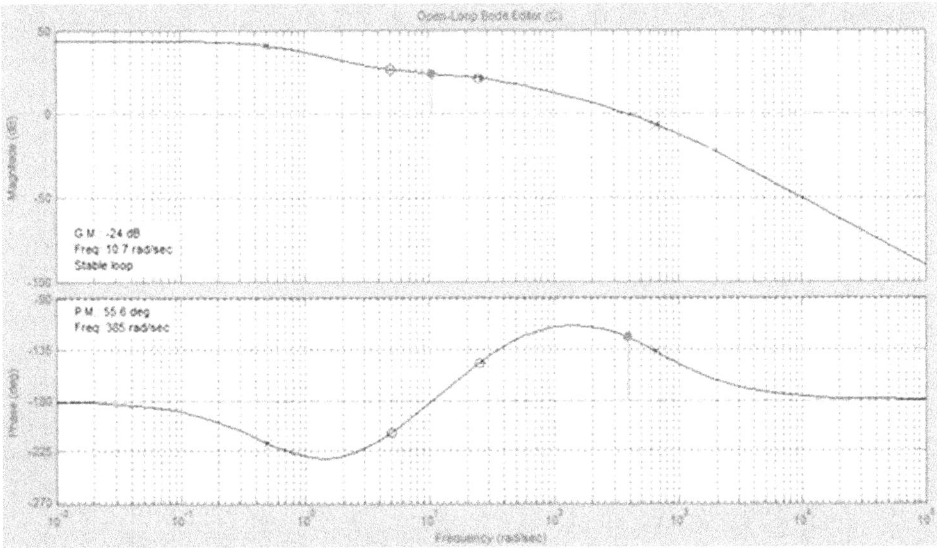

Figure 5. Position responses with lead-lag controllers at different operating points

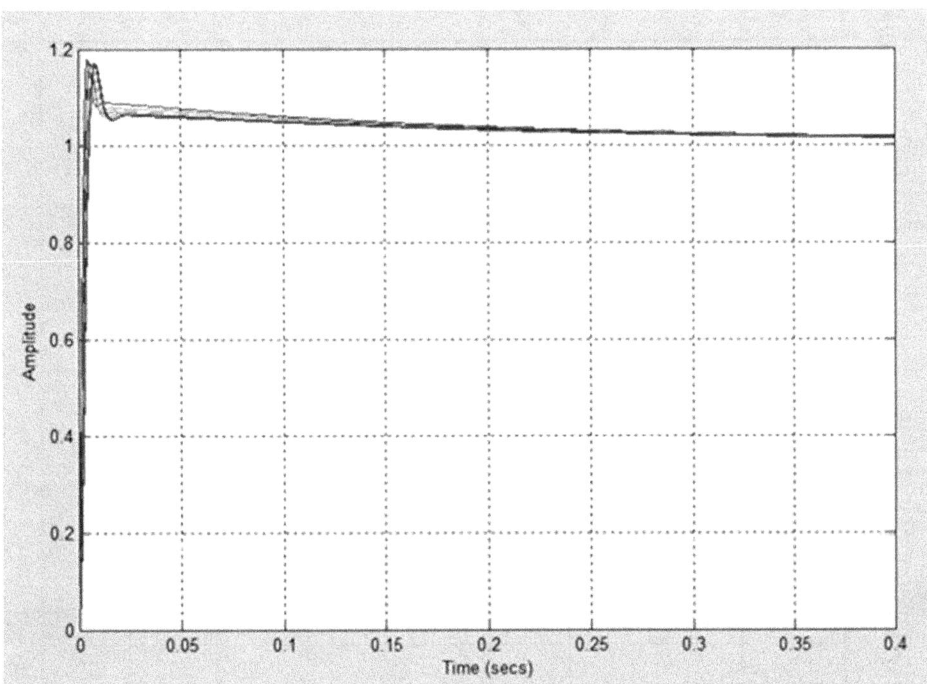

Table 2. Performances of lead-lag controller at different operating positions

Gap (mm)	T.F. of the Plant	TF of Lead-Lag	(%) Overshoot	Rise Time (sec)	Settling Time (sec)	S.S. Error	Margin of Stability (Kmin to Kmax)	Stable Closed Loop Poles
3	$\frac{45.74}{(s+59.22)(s-59.22)}$	$\left(19.1\left(\frac{s+57}{s+1180}\right)\left(\frac{s+5}{s+0.5}\right)\right)$	16.1	0.00218	0.277	0.01	1.58 to inf.	-5.45, -56.6, -559±699i
5	$\frac{25.32}{(s+39.62)(s-39.62)}$	$\left(19.8\left(\frac{s+37}{s+885}\right)\left(\frac{s+5}{s+0.5}\right)\right)$	16.1	0.00273	0.267	0.01	1.46 to inf.	-5.42, -36.5, -422±535i
7	$\frac{21.32}{(s+34.24)(s-34.24)}$	$\left(20.4\left(\frac{s+32}{s+805}\right)\left(\frac{s+5}{s+0.5}\right)\right)$	16.1	0.00286	0.253	0.01	1.35 to inf.	-5.39, -31.6, -384±508i
10	$\frac{13.98}{(s+26.97)(s-26.97)}$	$\left(21.2\left(\frac{s+25}{s+665}\right)\left(\frac{s+5}{s+0.5}\right)\right)$	16.1	0.00333	0.254	0.01	1.34 to inf.	-5.39, -24.7, -318±419i
12	$\frac{13.32}{(s+26.23)(s-26.23)}$	$\left(21.9\left(\frac{s+24}{s+655}\right)\left(\frac{s+5}{s+0.5}\right)\right)$	16.1	0.00334	0.253	0.01	1.36 to inf.	-5.39, -23.7, -313±418i
15	$\frac{12.37}{(s+25.60)(s-25.60)}$	$\left(22\left(\frac{s+23}{s+636}\right)\left(\frac{s+5}{s+0.5}\right)\right)$	16.1	0.00346	0.262	0.01	1.4 to inf.	-5.4, -22.6, -304±403i
17	$\frac{10.76}{(s+23.46)(s-23.46)}$	$\left(22.7\left(\frac{s+21}{s+597}\right)\left(\frac{s+5}{s+0.5}\right)\right)$	16.1	0.00359	0.258	0.01	1.39 to inf.	-5.4, -20.6, -286±384i
20	$\frac{10.04}{(s+21.98)(s-21.98)}$	$\left(23.1\left(\frac{s+20}{s+579}\right)\left(\frac{s+5}{s+0.5}\right)\right)$	16.1	0.00367	0.248	0.01	1.33 to inf.	-5.37, -19.7, -277±375i

Table 3. Performances of PID controller at different operating positions

Gap (mm)	T.F. of the Plant	T.F. of PID	(%) Overshoot	Rise Time (sec)	Settling Time (sec)	S.S. Error	Margin of Stability (Kmin to Kmax)	Stable Closed Loop Poles
3	$\frac{45.74}{(s+59.22)(s-59.22)}$	$0.0096\frac{(s+53)(s+5)}{s}$	16	0.00373	0.389	0	0.0015 to inf.	-720, -51.8, -5.52
5	$\frac{25.32}{(s+39.62)(s-39.62)}$	$0.0117\frac{(s+33)(s+5)}{s}$	16.1	0.00557	0.408	0	0.0019 to inf.	-260, -30.4, -6.19
7	$\frac{21.32}{(s+34.24)(s-34.24)}$	$0.012\frac{(s+28)(s+5)}{s}$	16.1	0.00643	0.415	0	0.0019 to inf.	-330, -26.4, -5.83
10	$\frac{13.98}{(s+26.97)(s-26.97)}$	$0.0145\frac{(s+21)(s+5)}{s}$	16.1	0.00813	0.433	0	0.0023 to inf.	-178, -18.4, -6.52
12	$\frac{13.32}{(s+26.23)(s-26.23)}$	$0.0149\frac{(s+20)(s+5)}{s}$	16	0.00834	0.438	0	0.0024 to inf.	-202, -17.7, -6.32
15	$\frac{12.37}{(s+25.60)(s-25.60)}$	$0.0157\frac{(s+19)(s+5)}{s}$	16.1	0.00855	0.444	0	0.0026 to inf.	-171, -16, -6.72
17	$\frac{10.76}{(s+23.46)(s-23.46)}$	$0.0165\frac{(s+17)(s+5)}{s}$	16.1	0.0097	0.454	0	0 0027 to inf.	-162, -14, -6.86
20	$\frac{10.04}{(s+21.98)(s-21.98)}$	$0.0167\frac{(s+16)(s+5)}{s}$	16	0.0102	0.457	0	0.0027 to inf.	-148, -12.9, -7.02

Figure 6. Position responses with PID controllers at different operating positions

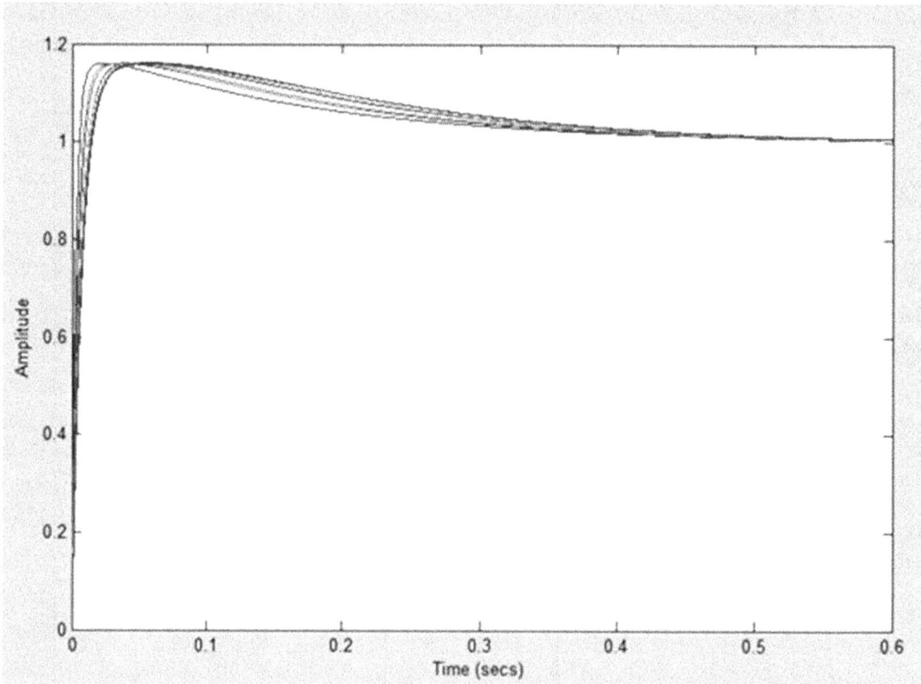

REALISATION OF CONTROL CIRCUIT

Since the basic control methodology of two independent unit used in the proposed electromagnetic levitation is similar, the hardware realization of the control circuit in two cases is also identical. Hence identical hardware control circuits have made for the two magnet-coils. The complete circuit diagram for one control card (for coil-1) is shown in Figure 9. Each control circuit consists of different parts such as generation of reference command (position gap) signal, position controller, current controller, PWM generation for the chopper switch and driver circuit for the MOSFET switch used in the chopper circuit. The reference gap command signal for the controller circuit is generated by a simple potential divider circuit that utilizes the available -15 volts supply along with a fixed resistor (1 K-ohm) and a variable resistor (1K-ohm pot) as shown in Figure 9. For noise free reference signal a capacitor (10μF, 35 volts electrolytic in parallel with 0.1μF ceramic) is connected across the potentiometer output.

The transfer function of the Lead compensator based on the theoretical design considering the overall system (without inner current loop) for an operating point of 10 mm gap is

$$G_{Lead}(s) = \frac{7.3(s+25)}{(s+323)} \tag{8.1}$$

The hardware realization of Lead compensator network can be done in many ways utilizing simple R-C combination with or without an operational amplifier. Here, an op-amp based Lead compensator (Fig.8.7) has been used. The pole location is significantly farther from the zero location and is chosen to reduce the effect of high frequency noise (due to EMI etc.) on the control circuit.

Figure 7. Circuit diagram of Lead controller

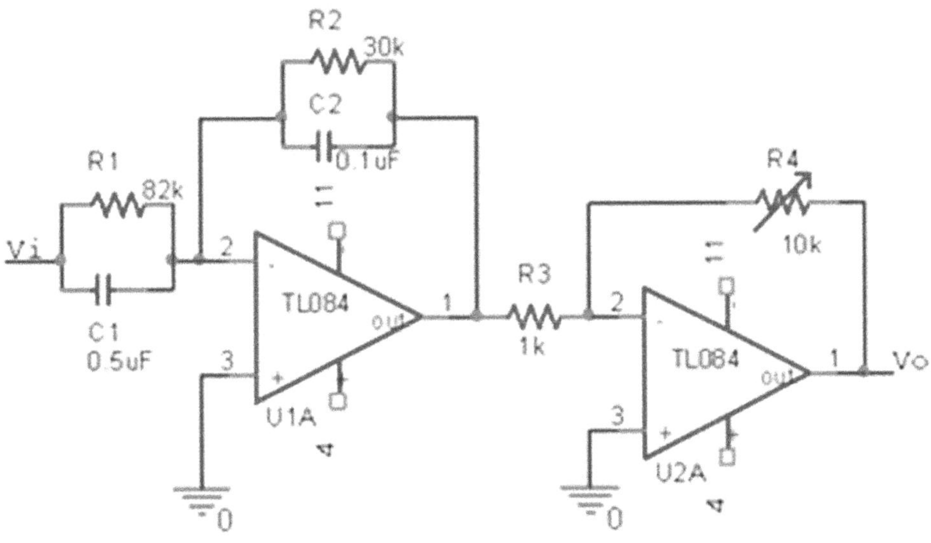

The transfer function of the overall Lead controller as shown in Figure 7 may be represented as

$$\frac{V_0(s)}{V_i(s)} = \frac{R_4 C_1 \left(s + \frac{1}{R_1 C_1} \right)}{R_3 C_2 \left(s + \frac{1}{R_2 C_2} \right)} \tag{8.2}$$

To realize the above transfer function by comparing Eqn.8.1 and Eqn.8.2, one gets, $\frac{1}{R_2 C_2} = 323$; Now choosing the value of C_2= 0.1μF, the value of R*2* becomes 31kΩ; Again for the zero of the compensator, $\frac{1}{R_1 C_1} = 25$; Here also the value of capacitor is taken as C_1= 0.5μF, so, R*1* is then found to be 80kΩ. For the implementation of dc gain of Lead compensator one more amplification stage is required. For the present case nearest standard values of resistors and capacitors have been considered.

The actual current is sensed by LEM make Hall-effect current sensor (LA-55P) and after scaling, it is fed to the differential amplifier circuit. The other input to the differential amplifier is the reference current signal generated from the position controller output. The current error signal output from the differential amplifier is processed through a PI controller. The current controller is realized by the following simple circuit. (Figure 8)

Figure 8. Circuit diagram of the current controller (PI controller)

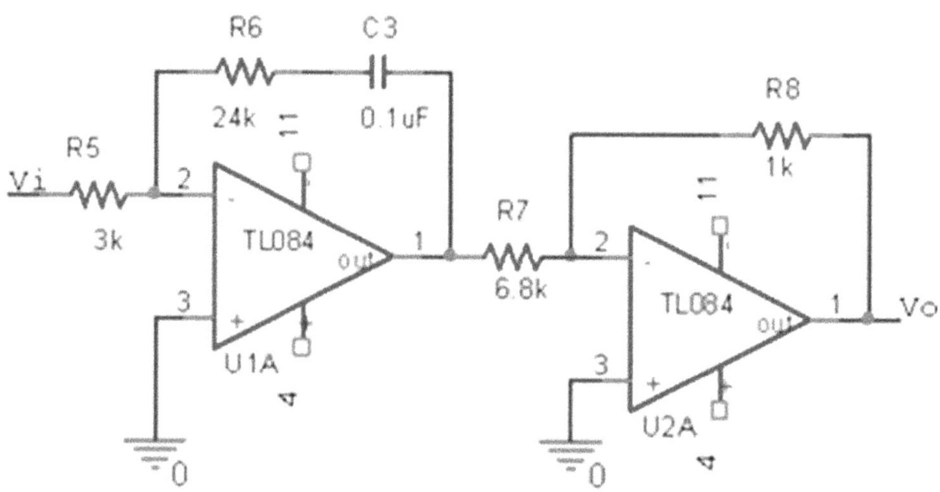

The transfer function of the current controller as shown in Figure 8 is given as

$$G_{cc}(s) = \frac{\left(R_6 + \dfrac{1}{sC_3}\right)}{R_5} = \left(\frac{R_6}{R_5} + \frac{1}{sR_5C_3}\right) \tag{8.3}$$

Comparing with the standard form of PI controller, the proportional gain, $K_p = \dfrac{R_6}{R_5}$ and the integral gain $K_i = \dfrac{1}{R_5C_3}$.

Now from Section the design value of proportional gain is, K_p= 8.1228 and integral gain is, K_i= 3194.5; To satisfy these values taking C_3= 0.1μF one gets, R5 $\cong$ 3kΩ and R6 $\cong$ 24kΩ;

Figure 9. Circuit diagram for overall control circuit (one unit)

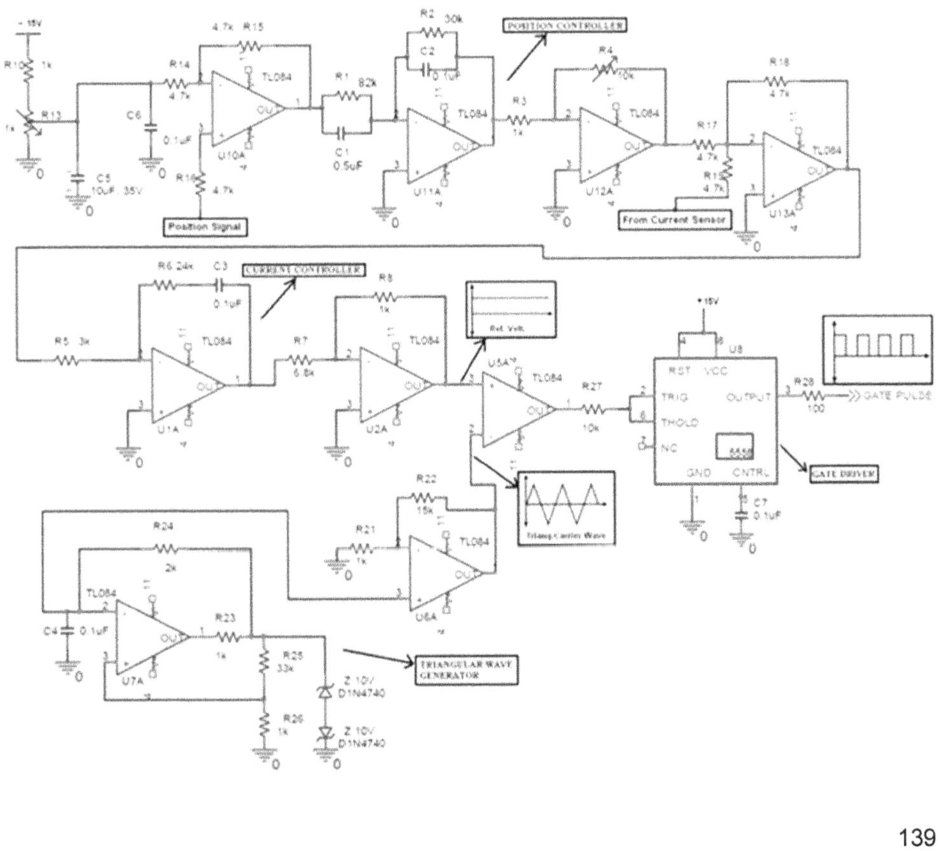

A control logic similar to the one described in Chapter 4 has been utilized to generate the PWM waveform for driving the MOSFET switches of the chopper circuit. To facilitate proper duty ratio control of the switch, a triangular wave is produced (described before). The output of the current controller is compared with the triangular wave for generation of the desired PWM waveform. The output of the comparator is not sufficient to drive the gate of the MOSFET switch and it is fed to an inverting buffer circuit using 555-IC. The detailed description of the gate drive circuit has been presented in Section 4.5.4 of Chapter 4.

POWER AMPLIFIER

Switched mode power circuits reported for the levitation purpose are mostly of H-type asymmetrical bridge topology shown below in Figure 10. It uses two switches and two diodes for single actuator system. This bridge circuit requires two isolated gate drives for the two switches. Some researchers have used linear power amplifiers for fast control of the magnet current (Carabelli, et. al. 2000; Maddaleno, et. al. 1997), but such amplifiers, when used for large current rated magnets, will be quite inefficient compared to switch mode circuits due to heavy power loss in the amplifier and the required volt-ampere rating of the amplifier transistor will also be many times higher. So from the energy efficiency point of view the H-type bridge circuit is more acceptable. But, the power circuit for the two-coil structure (Fig. 8.10) would require four switches and four diodes and this circuit would require four-isolated gate drives for the four switches. In effect the whole circuit not only becomes complicated but also becomes more cost effective. Since many hard switching devices are used in the circuit the possible EMI generation would also be more.

In this part of work two electromagnets with their individual air-gap controllers have been used for the levitation of the cylindrical rod. These magnets have their separate power amplifiers and they are separately controlled. Now instead of asymmetrical bridge circuit if the simplified circuit with only one switch (and a diode with energy-dump circuit) (Figure 11) per magnet is used, the overall saving in the power amplifier components may be significant. Two switches of the simplified power circuit may be connected such that their gate drive signals may not need ohmic isolation. Moreover two amplifiers may share a common energy dump capacitor-resistor circuit. The exact connection diagram is shown in Figure 11. This circuit uses half the number of switches and diodes used in the asymmetrical bridge circuit and as a result the total EMI generation is also less. The circuit is less energy-efficient compared to asymmetrical bridge circuit but more efficient compared to the linear amplifier.

In the power circuit when the switch is turned on, full supply voltage comes across the coil and the current in the coil increases. During the off period of the switch, the fast-recovery power diode gets forward biased. The capacitor voltage now comes across the coil and the coil voltage polarity reverses, facilitating quick decay of the coil current. As stated earlier, for stable suspension, one essential requirement of the power circuit is to quickly control the magnet current for fast control over the electromagnetic lift force. Thus, not only the current build up through the magnet coil needs to be fast, current decay must also be quite fast. The simple power circuit achieves this objective. Magnitude of discharge resistor across capacitor decides the mean value of capacitor voltage. The ripple in capacitor voltage is small due to high frequency operation. The coil-current also remains continuous and unidirectional due to high frequency chopping and the large inductance of the coil.

Figure 10. Schematic diagram of the asymmetrical bridge (H-bridge) converter circuit

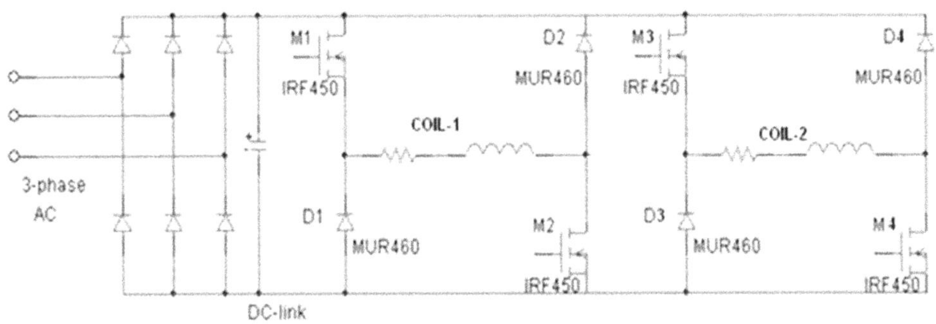

EXPERIMENTAL RESULTS

The main objective of this work is the stable suspension of the cylindrical rod at the desired operating air-gap position. The complete hardware set-up (Figure 12) has been fabricated in the laboratory and stable levitation of the cylindrical rod around 10 mm has been demonstrated in Figure 13. The advantage of the proposed system is the decoupled control of air-gap. Either through reference gap control or, change in controller parameters, any side of the rod may be controlled independently. Figure 14 shows demonstration of better pitching movement with decoupled control technique of each end of cylindrical rod. Here the reference gap of the left end (from front side) is less compare to right end of the levitated rod.

Figure 11. Simplified chopper circuit for two actuator based levitation system

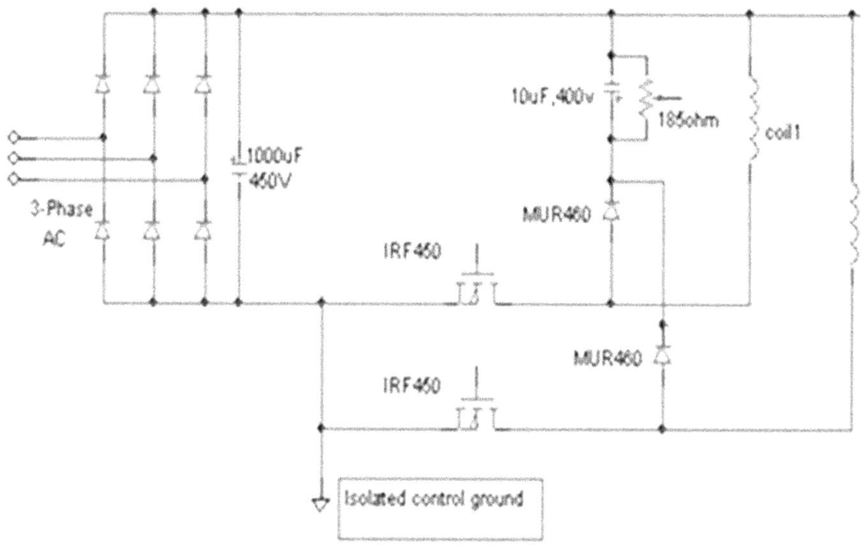

Figure 15 shows gate-pulses and coil-currents of both the systems during stable levitation of the ferromagnetic cylindrical rod. When the switches are turned on, full supply voltage appears across the coils and the currents in the coil increases. Thus energy flows from source to the magnet-coil. Duration of current flow through the coils in each switching cycle is equal to the ON duration of the gate driving pulses.

During ON time the energy dump capacitor (10μF, 400V) that holds some charge due to previous operation, discharges partially through the rheostat (185Ω). During OFF period of the switch, the coil currents find a path through fast-recovery power diodes to the resistor capacitor combination. The capacitor voltage effectively appears across the coils with a negative polarity (reverse of the polarity during ON duration).

Implementation of the Double Actuator-Based DCALS

Figure 12. Complete hardware set-up for the two actuator based DC attraction type levitation system

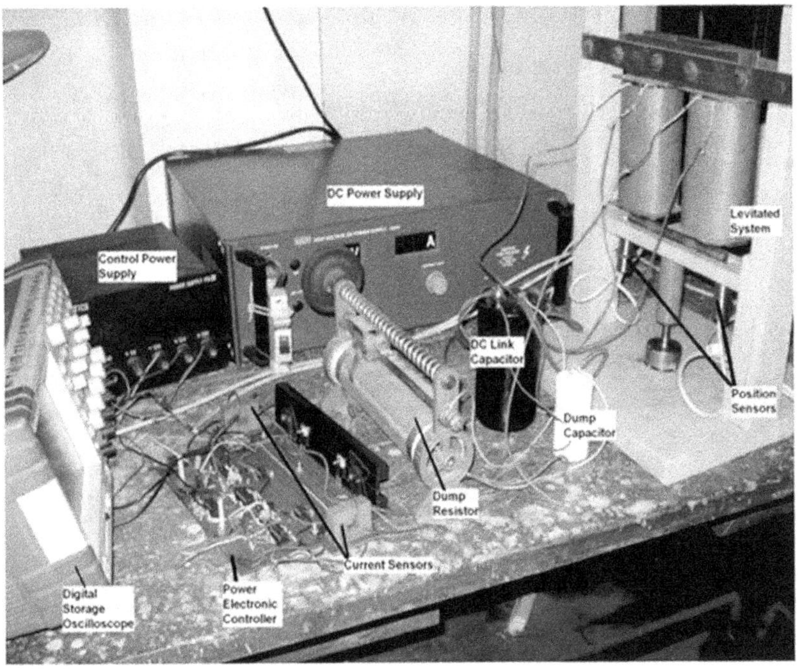

Figure 13. Photograph of the levitated rod at 10 mm air-gap

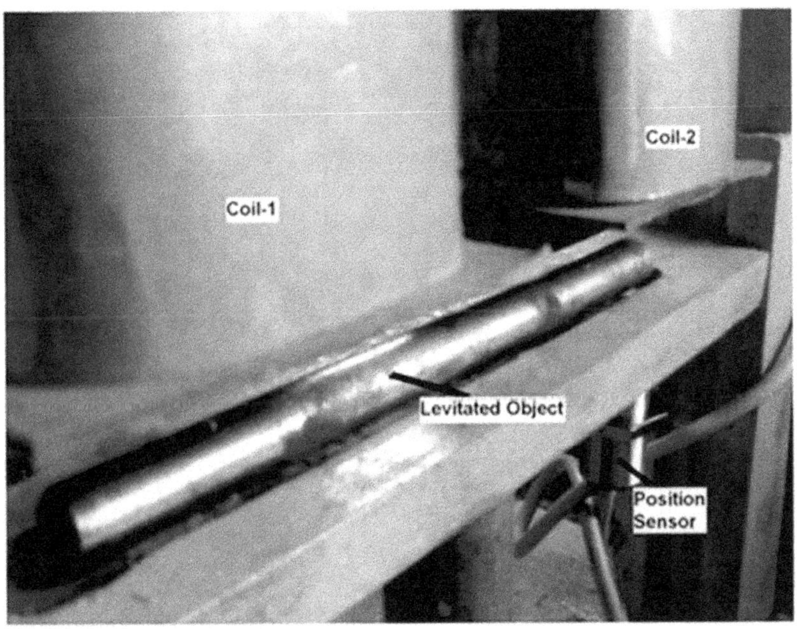

Figure 14. Demonstration of better pitching movement with decoupled control of the rod

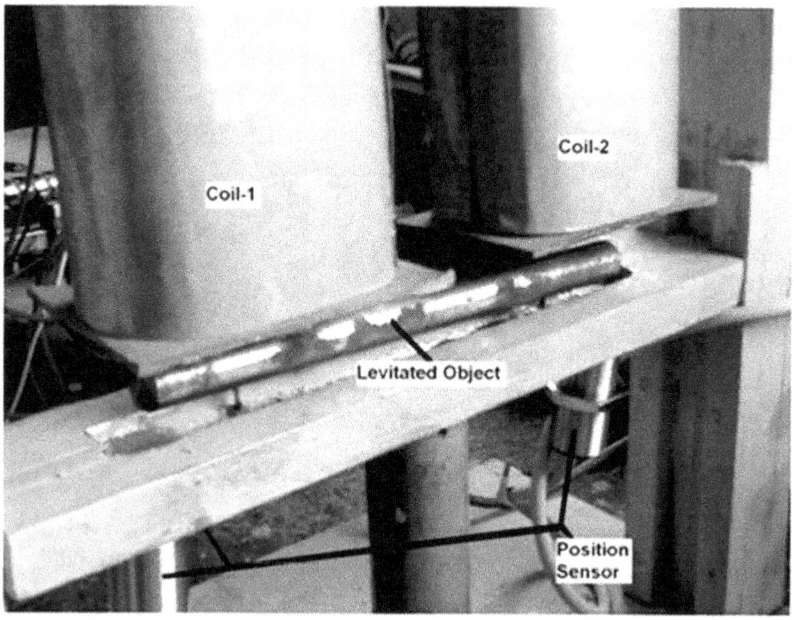

Figure 15. Gate pulses (Ch. 1 & Ch. 2) and coil currents (Ch. 3 & Ch. 4) during stable levitation of cylindrical rod

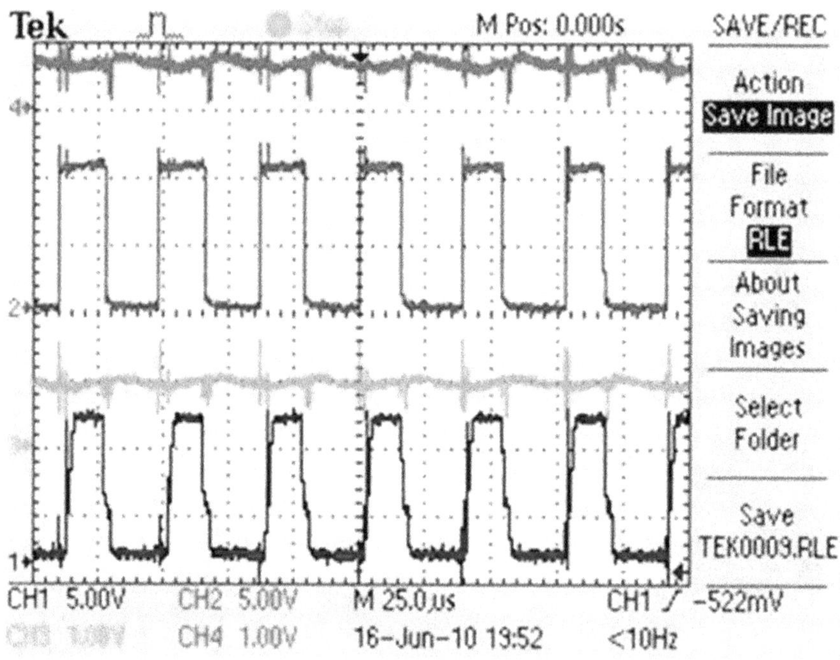

Figure 16. Coil voltages (Ch. 1 & Ch. 2) and coil currents (Ch. 3 & Ch. 4) during stable levitation of cylindrical rod

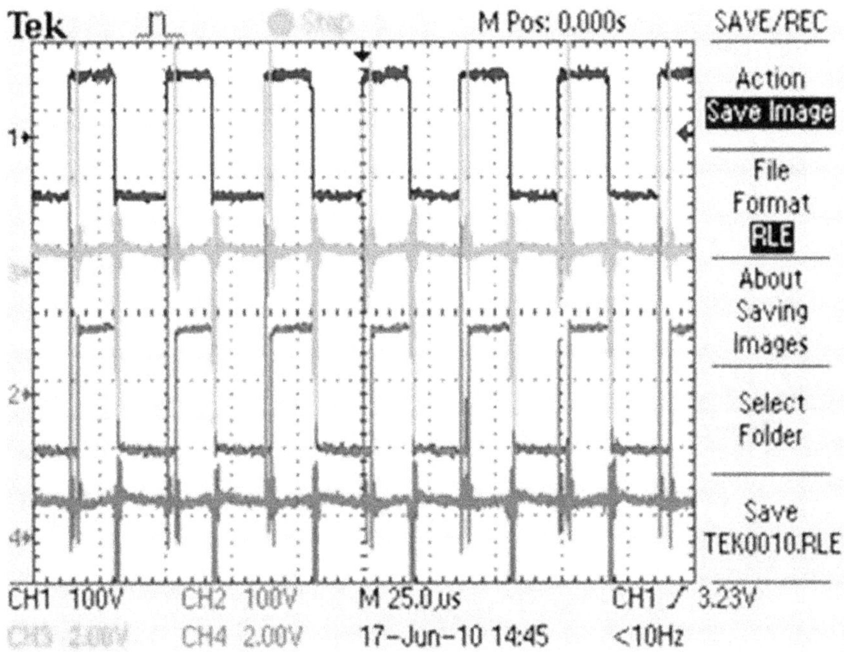

The two coil-voltages and two coil currents during stable suspension of the rod is shown in Figure 16. It is seen from the Figure 16 that both the actuators are excited by same amount of voltage (100 volt) and current (1 Amp) during stable levitation of the object at desired operating air-gap (10mm).

Figure 17 shows two coil voltages along with gate pulses of two switches during stable levitation of the object. The coil voltage is made equal positive and negative supply voltage (100 volts) with the predefined circuit parameters (resistor and capacitor). During OFF period of the switch, the capacitor starts charging with the flow of coil-energy into it. When the switch turns on, it discharges through the rheostat, thus, causing loss of voltage. Figure 18 shows coil-voltage, capacitor voltage, coil current and gate-pulse of one unit (during stable levitation of rod) when the DC input supply is 100 volts and the 185 ohm rheostat is set at 170 ohms.

Figure 17. Gate pulses (Ch. 3 & Ch. 4) and coil voltages (Ch. 1 & Ch. 2) during stable levitation of cylindrical rod

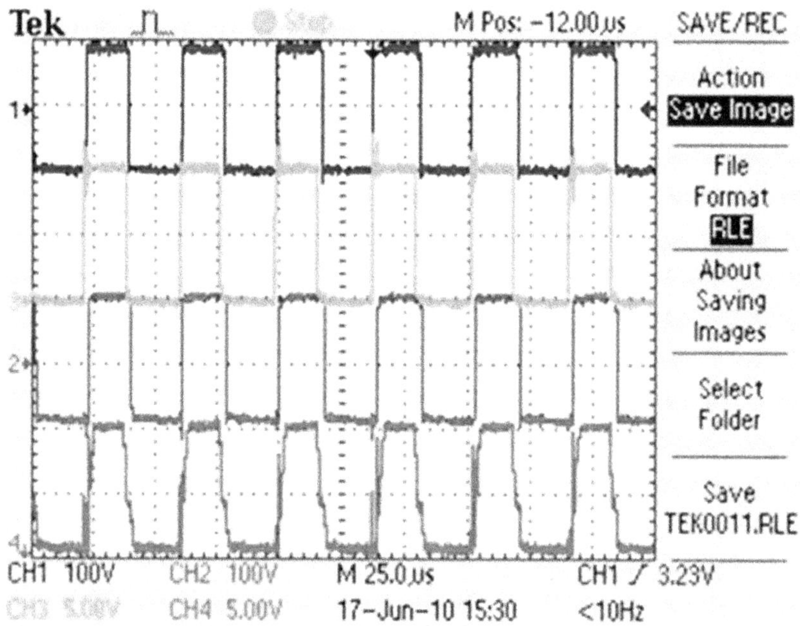

Figure 18. Coil voltage (Ch. 1), capacitor voltage (Ch. 2), coil current (Ch. 3), gate pulse (Ch. 4) during stable levitation of cylindrical rod

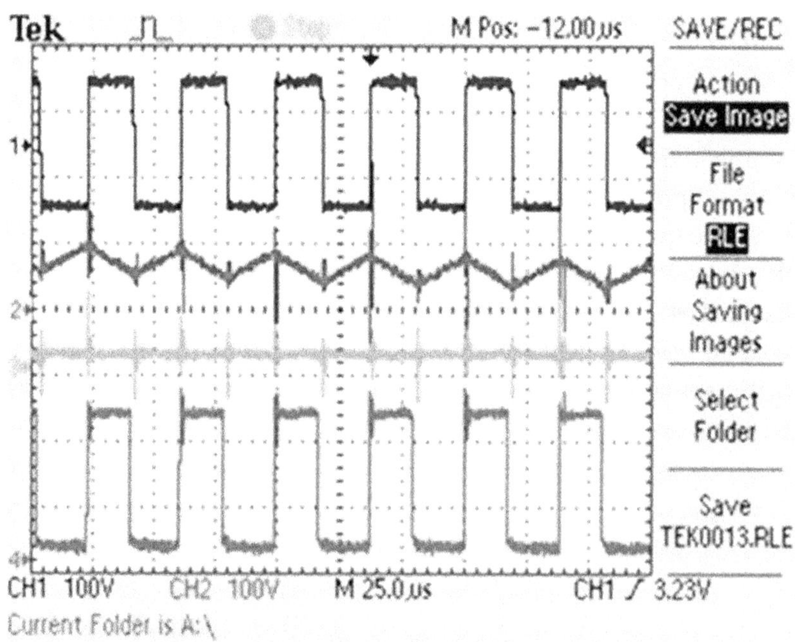

Figure 19. Two position signals (Ch. 1 & Ch. 2) and two current signals (Ch. 3 & Ch. 4) during stable levitation of cylindrical rod

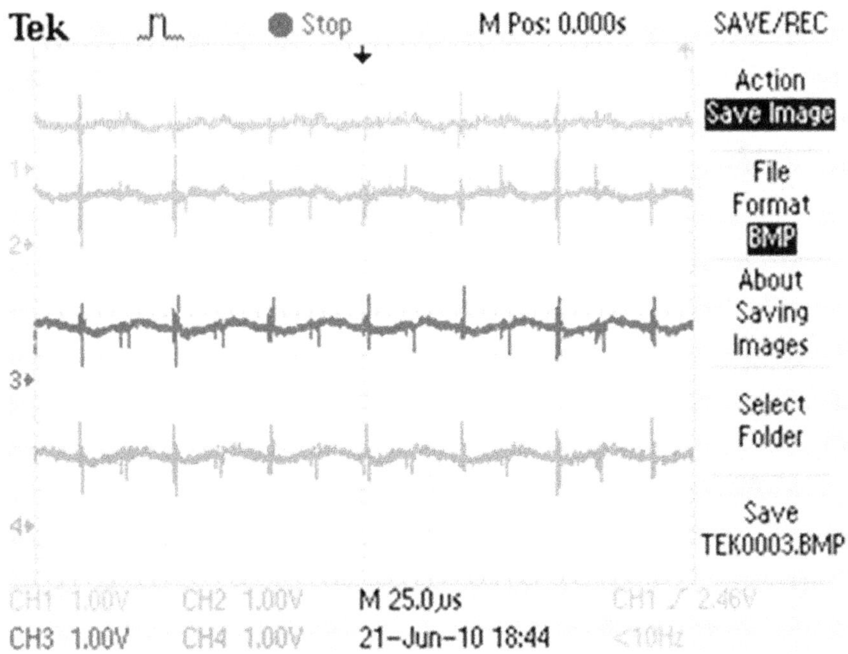

Figure 19 shows two positions and two current signals of the levitated system during stable levitation of the cylindrical rod at the desired operating air-gap . It is clear from the Fig.8.19 that both side of cylindrical rod has been lifted by same distance (around 10 mm shown in Figure 13) from the platform. The two acuators are drawing almost same amount of current from the controlled current source. The similar pattern of position and current signals indicates that the two systems are working in synchronism.

Figure 20 shows the dynamic position responses of two corners of rod. When the main switch of DC link power supply is suddenly ON, both end of rod gets vertical lift from the platform and the dynamic position responses for the two ends have been recorded. Similarly when the DC link power supply is switched OFF, the levitated rod falls due to gravity and this response is much faster than the response during lifting. This demonstration shows the stable movement control of the rod that lifts from a platform. The closed loop position responses are stable and have satisfactory transient response.

Figure 20. Dynamic position responses of two corners of rod

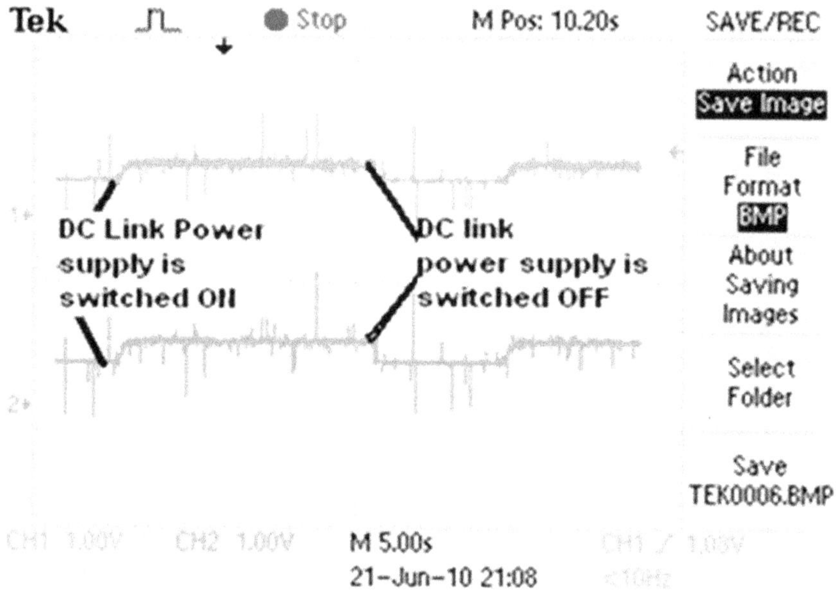

REFERENCES

Carabelli, S., Maddaleno, F., & Muzzarelli, M. (2000). High-efficiency linear power amplifier for active magnetic bearings. *IEEE Transactions on Industrial Electronics*, *47*(1), 17–24. doi:10.1109/41.824021

Maddaleno, F., Carabelli, S., & Muzzarelli, M. (1997). A modified class G amplifier for active magnetic bearings. *13th Annual Applied Power Electronics Conference Proceedings, 1*, 534–538.

Chapter 9
Optimization of Controller Parameters for the Proposed DCALS Using Genetic Algorithm (GA)

ABSTRACT

As presented in previous chapters, DCALS may be linearised for an operating air-gap, and the linearised system can be stabilized using conventional classical controllers. In the previous chapter, two electromagnet-based magnetic levitation systems have been demonstrated to suspend one cylindrical rod in the air. To achieve stable suspension of the rod, linearised modeling of the system has been developed at different operating air-gaps, and accordingly, some conventional classical controllers (lead, lead-lag, PID, PI plus lead) have been implemented. Thus, an optimization scheme for the proposed DCALS has been developed so that the prototype may stably operate over a wide area maintaining good performance.

INTRODUCTION

DC attraction type levitation system (DCALS) is inherently unstable and strongly non-linear in nature (Sinha, 1987; Jayawant, 1988). As presented in previous chapters, DCALS may be linearised for an operating air-gap and the linearised system can be stabilized using conventional classical controllers.

DOI: 10.4018/978-1-6684-7388-7.ch009

In the previous Chapter, two electromagnet based magnetic levitation system has been demonstrated to suspend one cylindrical rod in air. To achieve stable suspension of the rod, linearised modeling of the system has been developed at different operating air-gaps and accordingly some conventional classical controllers (Lead, Lead-Lag, PID, PI plus Lead) have been implemented. Resulting controllers have been tuned by extremely tedious and time consuming 'trial and error' process on the 's'-plane utilizing root-locus design concept. Here, the Ziegler-Nichols tuning rule for PID controller tuning, failed straight away since the plant has one zero in the RHS of 's'-plane. It has been found that these controllers provide satisfactory performance in the neighborhood of operating point and performance degradation occurs as the operating point is changed. It becomes difficult for the controllers to maintain stability if there is wide variation of operating air-gap. Due to the inherent nonlinear nature of the system, the uncertainty in the nominal linearised model increases as one move away from the nominal operating point. As a result, the controller becomes practically invalid for large variations of the operating air-gap. But from implementation point of view there is an urgent need to enhance the capability of the controller so that the prototype can stably levitate over a large air-gap variation. One possible solution that is mostly reported in the literature is the design of a non-linear controller by considering the nonlinear model of the system, and feedback linearizing controller (Joo, et. al. 1997; Trumper, et. al. 1997) is normally chosen. But finding the exact non-linear model of the maglev system as well as design and implementation of the controller is not an easy task.

Thus an optimization scheme for the proposed DCALS has to be developed so that the prototype may stably operate over a wide area maintaining good performance. For the tuning of controller parameters to get optimized performance two soft-computing techniques (i) Genetic Algorithm (GA) and (ii) Particle Swarm Optimisation (PSO) have been implemented in this part of work.

GA (Goldberg, 1989; Salami, 1995) and PSO (Kennedy, et. al. 1995; Hu, 2006) belonging particularly to the family of soft-computational algorithms which have been widely used in many control engineering applications. Soft-computation techniques have drawn much attention as optimization methods in the last two decades. From the optimization point of view, the main advantage of the aforesaid techniques is that they do not have much mathematical requirements about the optimization problems. All they need is an evaluation of the objective function. They are powerful optimization algorithms that work on a set of potential solutions, which is called population and swarm respectively. GA and PSO find the optimal solution through cooperation and competition among the potential solutions.

PSO shares many common points with GA. Both algorithms start with a group of a randomly generated population, both have fitness values to evaluate the population. Both update the population and search for the optimum with random techniques. Both systems do not guarantee success. However, PSO does not have genetic operators like crossover and mutation. Particles update themselves with the internal velocity. They also have memory, which is important to the algorithm. Compared with GAs, the information sharing mechanism in PSO is significantly different. In GAs, chromosomes share information with each other. So, the whole population moves like a one group towards an optimal area. In PSO, only gBest (global best) or, pBest (local best) gives out the information to others. It is a one way information sharing mechanism.

These algorithms are highly relevant for industrial applications, because they are capable of handling problems with nonlinear constraints, multiple objectives, and properties that frequently appear in real-world problems (Jan & Sulc, 2002).

GENETIC ALGORITHM (GA) OVERVIEW

GA is a subclass of evolutionary algorithms where the elements of the search space are strings or arrays of other elementary types. The concept of GA comes from the Darwinian idea of natural selection and survival of the fittest (Goldberg, 1989; Mahony, et. al. 2000). This technique was inspired by the mechanism of natural selection, a biological process in which stronger individuals are likely to be the winners in a competing environment. GA uses a direct analogy of such natural evolution to do optimization in order to solve highly complex problems.

The GA is composed of two main elements, which are strongly related to the problem being solved – the encoding scheme and the evaluation function (Sadasivarao, et. al. 2006; Salami, et. al. 1995). The encoding scheme is used to represent the possible solutions to the problem. It presumes that the potential solution of a problem is an individual and can be represented by a set of parameters. These parameters are regarded as the genes of a chromosome and can be structured by a string of concatenated values. The form of variables representation is defined by the encoding scheme. Individuals or, the variables can be encoded in some alphabets, binary strings, real numbers or, some other forms, depending on the application data. Its range, the search space, which is usually defined by the problem (Mahony, 2000; Shin, et.al. 2007) .

While applying GA practically, a population pool of chromosomes is installed and, initially, they are set to random values. In each cycle of genetic evolution, a subsequent generation is created from the chromosomes in the current population. The cycle of evolution is repeated until a termination criterion is reached (Shin, et.al. 2007). The number of evolution cycles, or a predefined fitness value can be set as the termination criterion.

Figure 1. Basic cycles of genetic algorithms

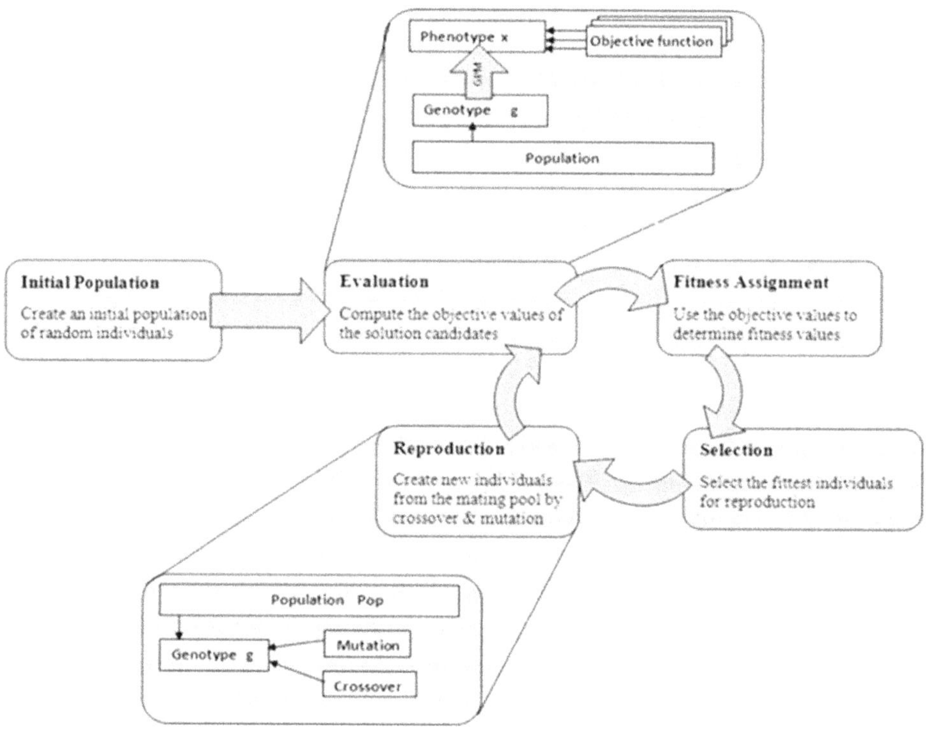

Since, the GA is a powerful search algorithm, hence an exploration of the search space can be performed which evolves in analogy to the evolution in nature (Rathikarani, et. al. 2007; Salami, et. al. 1995; Man, et. al. 1996; Kim, et. al. 2005). A GA (Figure 1) is typically initialised with a random population consisting of 20 to 100 individuals. This population (mating pool) is usually represented by real-valued numbers or, binary strings called chromosomes.

How well an individual performs a task is measured and assessed by the objective function. The fitness of each chromosome is assessed and a survival of the fittest strategy is applied. Each iteration of the algorithm is termed as a generation. The evolution of solutions is simulated through a objective function and genetic operators such as reproduction, crossover and mutation (Salami, et. al. 1995; Goldberg, 1989). The fittest individual will survive generation after generation (known as elitism), while also reproducing and generating offspring's that might be stronger. At the same time, the weakest individuals disappear from each generation. The flow chart is given in Figure 2 and the brief outline of the GA program used in this work is shown in the following steps.

Declaration of parameter matrices;
defining population and generation size;
defining the maglev process;
initializing random population;
for loop of generation starts;
for loop of population starts;
{ finding partial fraction expansion of the overall system TF;
evaluating chromosomes in population using objective function;
finding fittest solutions;
performing elitism by keeping the best chromosomes for the next generation;
selecting parents for reproduction operation;
performing crossover operation;
if necessary conducting mutation operation;
storing the new best solution;
new population = population + new best solution; }
end for;
end for;
displaying comparative plots and iteration plots.

Reproduction

During the reproduction phase each chromosome is assessed by the objective function to check its fitness. After evaluation it is required to create a new population from the current generation. The selection operation determines which parents participate in producing offspring for the next generation, and it is analogous to "survival of the fittest" in natural selection.

Figure 2. Flow chart for the GA programming

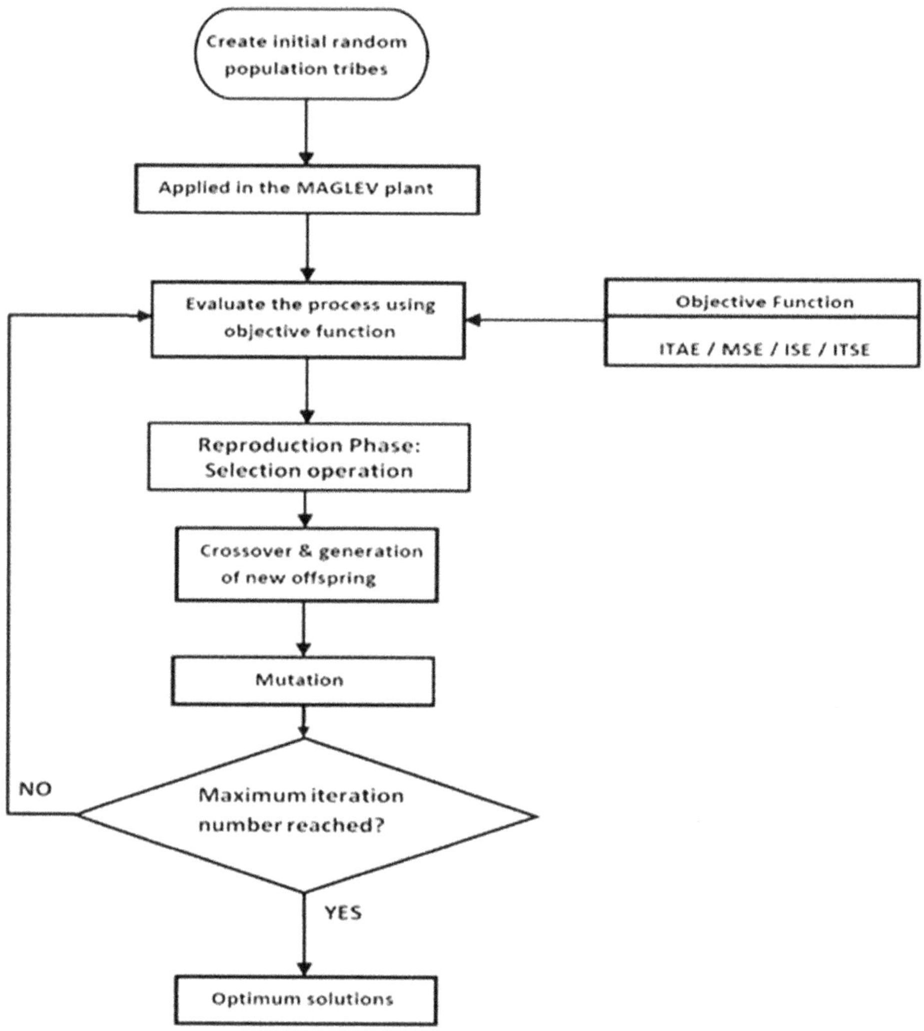

While to carry out selection process the fittest chromosomes are utilized for reproduction operation. Now this is just like natural evolution, where a fit chromosome has a higher probability of being selected for reproduction. The probability of an individual being selected is, thus, related ultimately to the objective function, ensuring that fitter individuals are more likely to leave offspring.

Crossover

To exploit the potential of the current gene pool, it is necessary to use crossover process to generate new chromosomes that is hoped of retaining good features from the previous generation. Crossover is usually applied to selected pairs of parents with a probability equal to a given crossover rate. New individuals are generally created as offspring of two parents. One or more so called crossover points are selected (usually at random) within the chromosome of each parent, at the same place in each. The parts delimited by the crossover points are then interchanged between the parents. The crossover operations swap certain parts of the two selected strings in order to capture the good parts of old chromosomes and create better new ones. The crossover probability indicates how often crossover is performed. The simplest crossover technique is the Single Point Crossover. In Figure 3 it is shown that how the crossover operation has been done in the present case.

Figure 3. Crossover operation

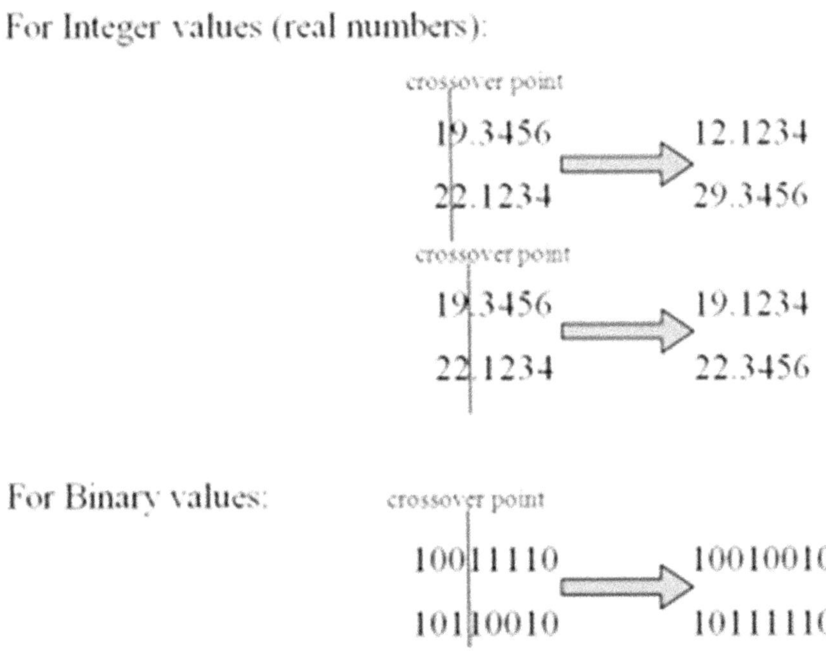

Mutation

Crossover operation exploits current gene potentials, but if the population does not contain the solution within the required solution area to optimize a particular problem, no amount of gene mixing can produce a satisfactory solution. For this reason, a mutation operation capable of spontaneously generating new chromosomes is included. The most common way of implementing mutation is to flip a bit of binary number or, abruptly changing any integer value with a probability equal to a very low given mutation rate. A mutation operation can prevent any single bit from converging to a value throughout the entire population and, more important, it can prevent the population from converging and stagnating at any local optima. The mutation rate is usually kept low, so, good chromosomes obtained from crossover are not lost.

OPTIMIZATION OF CONTROLLER PARAMETERS USING GA

In this work, the main objective is to develop an optimized controller which can provide satisfactory performance irrespective of any operating zone for the proposed DCALS (Chapter 6). The parameters of classically designed controller at nominal operating air-gap (10 mm) have been optimized utilizing GA and a comparative study has been made between the dynamic performances of classically designed and the optimized controllers at nominal as well as other off-nominal operating air-gaps (low and high).

The parameters of the Lead, Lead-Lag and PID controllers (that had been designed in Chapter 6) are to be optimized using GA. There are three parameters for Lead controller, such as, one DC gain, one pole and one zero and for PID controller the parameters are, one DC gain and two zeros but, for Lead-Lag controller there are five parameters namely, one DC gain, two zeros & two poles.

From Table 1 (Chapter 6) the linearized unstable maglev plant may be designated as $G_p^{(j)}(s)$: where,

$$G_p^{(1)}(s) = \frac{45.74}{(s+59.22)(s-59.22)} \text{ for 3mm air gap (low gap zone) at j=1;} \tag{9.1}$$

$G_p^{(2)}(s) = \dfrac{13.98}{(s+26.97)(s-26.97)}$ for 10mm air gap (medium gap zone) at j=2; (9.2)

$G_p^{(3)}(s) = \dfrac{10.76}{(s+23.46)(s-23.46)}$ for 17mm air gap (higher gap zone) at j=3; (9.3)

It is desired to stabilize $G_p^{(j)}(s)$, j=1,2,3 by the nominal controller $C^{(k)}(s)$, k = 1,2,3 (k is index to denote the three different controllers namely Lead, Lead-Lag and PID designed at nominal operating air-gap of 10 mm) using only output (Man, et. al. 1996; Kim, et. al. 2005; Gaing, 2004; Krohling, 1997) feedback (Figure 4)

Figure 4. Block diagram of simplified DCALS

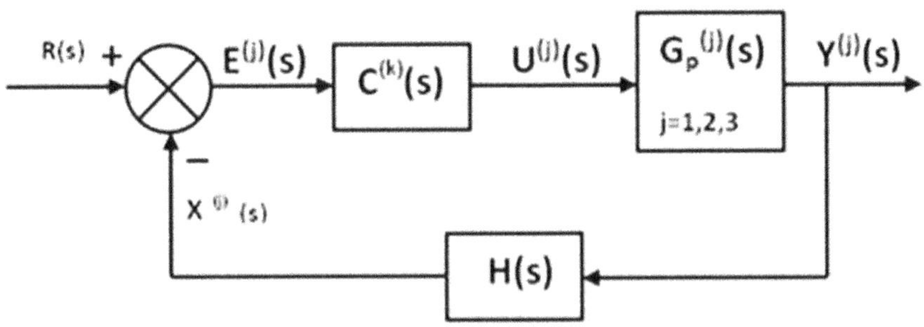

The general structure of Lead controller is $C^{(K)}(s) = \dfrac{K^{(j)}(s+z_c^{(j)})}{\left(s+p_c^{(j)}\right)}$ for stabilizing $G_p^{(j)}(s)$, for j=1,2,3 and k=1.

For the Lead controller, the DC gain is populated in the range of 4 to 8, similarly, the possible solution area for zeros and poles are chosen in the ranges of 10 to 25 and 280 to 380 respectively. This is nothing but the area of seeding for the possible optimum solutions.

In a similar manner, for the Lead-Lag controller, operating DC gain is populated in the range of 20 to 70, Lead-zeros and Lead-poles of the Lead-Lag controller are kept in the ranges of 10 to 25 and 500 to 2000 respectively and

Lag-poles and Lag-zeros of the Lead-Lag controller are kept in the ranges of 0.3 to 0.6 and 3 to 6 respectively.

For the PID controller operating DC gain has been populated in the range of 0.005 to 0.05, the zero near to origin is populated in range of 1 to 7 and the zero near to the stable pole of maglev plant is kept in the range of 9 to 25.97.

Above mentioned ranges are conceived from the values of parameters of classically designed Lead, Lead-Lag and PID controllers as described in Chapter 6 (Table 3, Table 4 and Table 5).

The general structure of the Lead-Lag controller is

$$C^{(K)}(s) = \frac{K^{(j)}(s + z_{c1}^{(j)})(s + z_{c2}^{(j)})}{(s + p_{c1}^{(j)})(s + p_{c2}^{(j)})}$$

for stabilizing $G_p^{(j)}(s)$, for j=1,2,3 and k=2.

In the present work one simplified structure of PID controller has been considered and is given by:

$$C^{(j)}(s) = \frac{K^{(j)}(s + z_{c1}^{(j)})(s + z_{c2}^{(j)})}{s}$$

for stabilizing $G_p^{(j)}(s)$, for j=1,2,3 and k=3.

The three nominal classical controllers $C^{(k)}(s)$, k=1,2,3 are as follows:

$$C^{(1)}(s) = \left(7.3\frac{s+25}{s+323}\right)$$

is the nominal Lead controller for k=1; (9.4)

$$C^{(2)}(s) = \left(21.2\left(\frac{s+25}{s+665}\right)\left(\frac{s+5}{s+0.5}\right)\right)$$

is the nominal Lead-Lag controller for k=2; (9.5)

$$C^{(3)}(s) = \left(0.0145\frac{(s+21)(s+5)}{s}\right)$$

is the nominal PID controller for k=3; (9.6)

For designing proposed controllers it is required to define & initialize population size, length of chromosome, number of iterations, selection, crossover and mutation types, crossover rate etc. Selection of these parameters decide, to a great extent, the ability of the designed controller ((Man, et. al. 1996; Kim, et. al. 2005)

For all the proposed controllers the required declarations inside the GA program are as follows:

- Population size: 50
- Number of generations: 100
- Crossover type: Single point crossover
- Crossover point: Randomly generated
- Crossover rate: 1
- Mutation type: Uniform

DEFINING OBJECTIVE FUNCTION FOR GA

To evaluate fitness of each chromosome the objective function is implemented, and choosing the objective function is the most crucial step for application of GA. Sometimes performance indices are used as the objective functions. Several available performance indices are Mean of the Squared Error (MSE), Integral of Time multiplied by Absolute Error (ITAE), Integral of the Squared Error (ISE), Integral of Time multiplied Squared Error (ITSE) and Integral of Absolute Magnitude of the Error (IAE) etc. (Gundogdu, 2005; Gaing, 2004)

$$\text{MSE} = \frac{1}{t}\int_0^T \left(\mathrm{e(t)}\right)^2 \mathrm{dt}\,, \ \text{IAE} = \int_0^T \left|\mathrm{e(t)}\right| \mathrm{dt}\,,$$

$$\text{ITAE} = \int_0^T t\left|\mathrm{e(t)}\right| \mathrm{dt}\,, \ \text{ISE} = \int_0^T \left(\mathrm{e(t)}\right)^2 \mathrm{dt}\,, \ \text{ITSE} = \int_0^T t\left(\mathrm{e(t)}\right)^2 \mathrm{dt}\,.$$

In this work the performance of a controller is evaluated in terms of ITAE criteria. The time limits are from $t = 0$ to $t = \tau$, where $\tau = T_s$ is the settling time of the system to reach steady state condition for a unit step input. Here, the value of settling time (T_s) are considered as; 0.0281 sec, 0.254 sec and 0.433 sec for Lead, Lead-Lag, and PID controller respectively.

To find out the error function e(t) of the proposed system, at first, it is required to find out the 's'-domain polynomial and then transformation into time-domain is to be done. The error signal E(s) is produced for each member of the population in each generation cycle. From Figure 4, the overall closed loop transfer-function becomes:

$$\frac{Y^{(j)}(s)}{R(s)} = \frac{C^{(k)}(s)G_p^{(j)}(s)}{1+C^{(k)}(s)G_p^{(j)}(s)H(s)}, \tag{9.7}$$

$$\Rightarrow Y^{(j)}(s) = \frac{C^{(k)}(s)G_p^{(j)}(s)}{1+C^{(k)}(s)G_p^{(j)}(s)H(s)}R(s) \tag{9.8}$$

And the error function can be written, $E^{(j)}(s) = R(s) - X^{(j)}(s)$

$$\Rightarrow E^{(j)}(s) = R(s) - H(s)Y^{(j)}(s) \tag{9.9}$$

By (9.9) from (9.10),

$$\Rightarrow E^{(j)}(s) = R(s) - H(s)\frac{C^{(k)}(s)G_p^{(j)}(s)}{1+C^{(k)}(s)G_p^{(j)}(s)H(s)}R(s)$$

$$\Rightarrow E^{(j)}(s) = \frac{R(s)}{1+C^{(k)}(s)G_p^{(j)}(s)H(s)} \tag{9.10}$$

In case of Lead controller, for k=1, $C^{(1)}(s) = K^{(1)}\frac{(s+z_c^{(1)})}{(s+p_c^{(1)})}$ & plant

$G_p^{(j)}(s) = \frac{a^{(j)}}{(s^2 - b^{(j)})}$, as $a^{(j)} = \frac{K_i^{(j)}}{m}, b^{(j)} = \frac{K_z^{(j)}}{m}$

H(s) = c (constant), also, for unit step input R(s) = $\frac{1}{s}$,

Therefore the Eqn.9.10 is becoming,

$$\Rightarrow E^{(j)}(s) = \frac{(s + p_c^{(1)})(s^2 - b^{(j)})}{s(s + p_c^{(1)})(s^2 - b^{(j)}) + sK^{(1)}(s + z_c^{(1)})a^{(j)}c} \tag{9.11}$$

In the present work, partial fraction expansion approach is used to obtain time-domain polynomial, which is rather less time consuming and lesser error prone. In MATLAB partial fraction expansion for the computed transfer functions are done by

[r, p, k] = residue (N, D)

where, r, p and k will produce residues, poles and gains respectively.

For distinct real roots the time domain polynomial function for error can be defined generally as

$$e(t) = r(1) * e^{p(1)t} + r(2) * e^{p(2)t} + r(3) * e^{p(3)t} \tag{9.12}$$

For repeated roots at origin the error time domain polynomial function can be defined generally as

$$e(t) = r(1) * e^{-p(1)t} - r(2) + r(3) * t \tag{9.13}$$

For complex conjugate roots the error time domain polynomial function is defined generally as

$$e(t) = M * e^{-\alpha\, t} * \cos(\omega t + \varphi) + r(3) + r(4) * t \tag{9.14}$$

where, M=2*|r(1)|, φ=*angle*(r(1)), ω=*image*(*p*(*1*)), α=re*al*(*p*(1)).

After choosing the appropriate polynomial the error in time-domain is computed and minimization of the objective function is performed in MATLAB environment (Cadence Design Systems, 2000).

TERMINATION CRITERION FOR GA

Normally, the termination criterion of the optimization algorithm may be either achieving maximum number of iterations or, the attainment of satisfactory objective function value. In this work the termination criteria has been considered to be the maximum number of iterations (or, generations) at 100.

RESULTS OBTAINED FROM GA OPTIMIZATION

It is aimed to optimize the parameters for classical Lead, Lead-Lag and PID controller designed at 10 mm operating air-gap (nominal gap) of the proposed DCALS utilizing GA. The performance of all the optimized controllers has to be tested in nominal as well as off-nominal air-gaps. The parameters of all these controllers have been considered after 100 iterations in the optimization process. It is observed (Figure 9 to Figure 11) that the parameter values are converging nicely towards the near optimal area.

The three optimized controllers $C^{(k)}(s)$, k=1,2,3 are as follows (referring Tables 2, 4 and 6 in Chapter 7)

$$C^{(1)}(s) = \left(4.0109\frac{s+24.9409}{s+349.4228}\right)$$

is Lead controller for k=1; (9.15)

$$C^{(2)}(s) = \left(68.949\left(\frac{s+24.3039}{s+1922.1}\right)\left(\frac{s+3.2397}{s+0.5611}\right)\right)$$

is Lead-Lag controller for k=2; (9.16)

$$C^{(3)}(s) = \left(0.04688\frac{(s+25.8638)(s+6.9794)}{s}\right)$$

is PID controller for k=3; (9.17)

Now, the GA based near optimal nominal gap controllers have been operated at low (3 mm), nominal (10 mm) and high gap (17 mm). The performances of all the optimized controllers at three different air-gaps have been observed and a comparative study with the performances of classically designed controllers operating at similar air-gaps has been made.

Figure 5 and Table 1 shows comparative transient position responses of DCALS with classical and GA based nominal Lead controller while operating at nominal and other off-nominal (low and high) air-gaps. When the nominal classical Lead controller is operated at nominal air-gap (10 mm), it shows 16% overshoot with rise time of 0.00536 sec, but with the GA based nominal controller, the percentage overshoot of nominal air-gap position response is 0.443% with rise time 0.0111 sec. Similarly, when the nominal classical controller is operated at lower gap (3 mm), the percentage overshoot increased to a value of 27.5% with rise time of 0.00252 sec, but the GA based controller reduces the overshoot to 9.6% with rise time of 0.00563 sec. Again when the nominal classical controller is made to operate at high (17 mm) operating air-gap, then overshoot in position response is 13.2% with rise time of 0.00653 sec, but for GA based controller it reduces to 2.26% with rise time of 0.014 sec (detailed values are shown in Table 1). It is to be mentioned here that for Lead compensator steady-state error cannot be minimized while reducing the overshoot simultaneously. This problem of non-minimization in the steady-state error with the Lead controllers eventually tends to slower position responses. The parameters for classical and GA based nominal gap Lead controller is shown in Table 2.

Figure 5. Comparative position responses of DCALS with classical and GA based nominal Lead controller operating at nominal and off-nominal (low and high) air-gaps

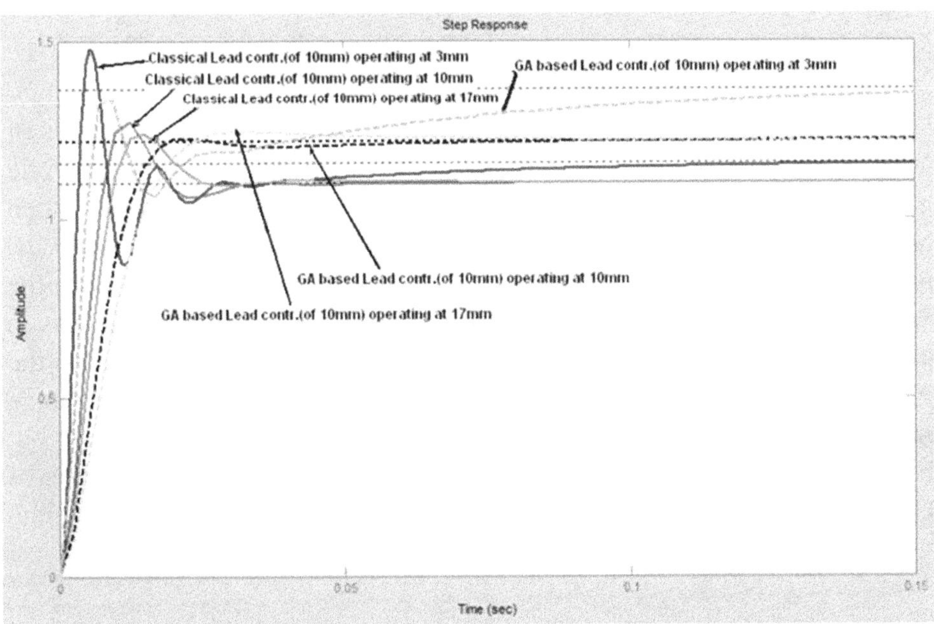

Table 1. Comparative dynamic performances of DCALS with classical and GA based nominal lead controller while operating at nominal and other off-nominal (low and high) air-gaps

Air Gap	Specifications	Classical	GA
	Peak Overshoot(%)	27.5	9.6
	Settling Time(sec)	0.0767	0.0527
3mm (low)	Rise Time(sec)	0.00252	0.00563
	S.S.Error	0.16	0.37
	Peak Overshoot(%)	16	0.443
	Settling Time(sec)	0.0281	0.0172
10mm (nominal)	Rise Time(sec)	0.00536	0.0111
	S.S.Error	0.1	0.22
	Peak Overshoot(%)	13.2	2.26
	Settling Time(sec)	0.0229	0.0381
17mm	Rise Time(sec)	0.00653	0.014
(high)	S.S.Error	0.1	0.22

Table 2. Comparative parameters of lead controller at 10mm air-gap

Parameters	Classical	GA
Gain (K)	7.3	4.0109
Zero (Z)	25	24.9409
Pole (P)	323	349.4228

Figure 6 and Table 3 shows comparative transient position responses of DCALS with classical and GA based nominal Lead-Lag controller while operating at nominal and other off-nominal (low and high) air-gaps. When the nominal classical Lead-Lag controller is operated at nominal air-gap (10 mm), it shows 16.1% overshoot with rise time of 0.00333 sec, but with the GA based nominal controller, the percentage overshoot of nominal air-gap position response is 4.65% with rise time 0.00304 sec. Similarly, when the nominal classical controller is operated at lower gap (3 mm), the percentage overshoot increased to a value of 34.4% with rise time of 0.0017 sec, but the GA based controller reduces the overshoot to 8.97% with rise time of 0.00124 sec. Again when the nominal classical controller is made to operate at high (17 mm) operating air-gap, then overshoot in position response is 13.5% with rise time of 0.0041 sec, but for GA based controller it reduces to 5.84% with rise time of 0.00411 sec (detailed values are shown in Table 3). The parameters for classical and GA based nominal gap Lead-Lag controller is shown in Table 4.

Table 3. Comparative dynamic performances of DCALS with classical and GA based nominal Lead-Lag controller operating at nominal and other off-nominal (low and high) air-gaps

Air Gap	Specifications	Classical	GA
	Peak Overshoot(%)	34.4	8.97
	Settling Time(sec)	0.332	0.443
3mm (low)	Rise Time(sec)	0.0017	0.00124
	S.S.Error	0.01	0.01
	Peak Overshoot(%)	16.1	4.65
	Settling Time(sec)	0.254	0.338
10mm (nominal)	Rise Time(sec)	0.00333	0.00304
	S.S.Error	0.01	0.01
	Peak Overshoot(%)	13.5	5.84
	Settling Time(sec)	0.248	0.331
17mm (high)	Rise Time(sec)	0.0041	0.0040
	S.S.Error	0.01	0.01

Table 4. Comparative parameters of lead-lag controller at 10mm air-gap

Parameters	Classical	GA
Gain (K)	21.2	68.949
Lead Zero (Z1)	25	24.3039
Lead Pole (P1)	665	1922.1
Lag Zero (Z2)	5	3.2397
Lag Pole (P2)	0.5	0.5611

Figure 6. Comparative position responses of DCLAS with classical and GA based nominal lead-lag controller operating at nominal and other off-nominal (low and high) air-gaps

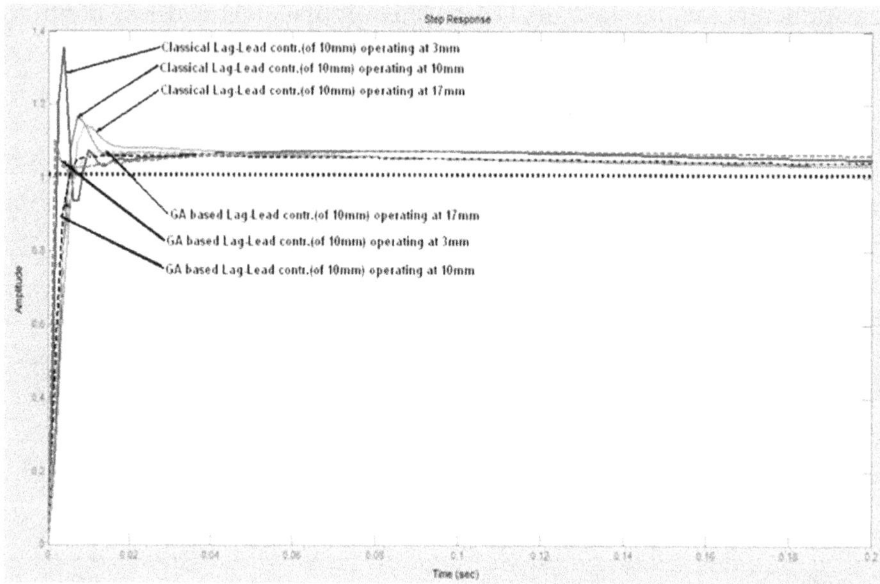

Similarly, Figure 7 and Table 5 shows comparative transient position responses of DCALS with classical and GA based nominal PID controller while operating at nominal and other off-nominal (low and high) air-gaps. When the nominal classical PID controller is operated at nominal air-gap (10 mm), it shows 16.1% overshoot with rise time of 0.00813 sec, but with the GA based nominal controller, the percentage overshoot of nominal air-gap position response is 4.96% with rise time 0.00299 sec. Similarly, when the nominal classical controller is operated at lower gap (3 mm), the percentage overshoot increased to a value of 19.7% with rise time of 0.003 sec, but the GA based controller reduces the overshoot to 4.3% with rise time of 0.00116 sec. Again when the nominal classical controller is made to operate for high (17 mm) operating air-gap, then overshoot in position response is 18.3% with rise time of 0.00992 sec, but for GA based controller it reduces to 6.11% with rise time of 0.00372 sec (detailed values are shown in Table 5). The parameters for classical and GA based nominal gap PID controller is shown in Table 6.

Figure 7. Comparative position responses of DCALS with classical and GA based nominal PID controller operating at nominal and other off-nominal (low and high) air-gaps

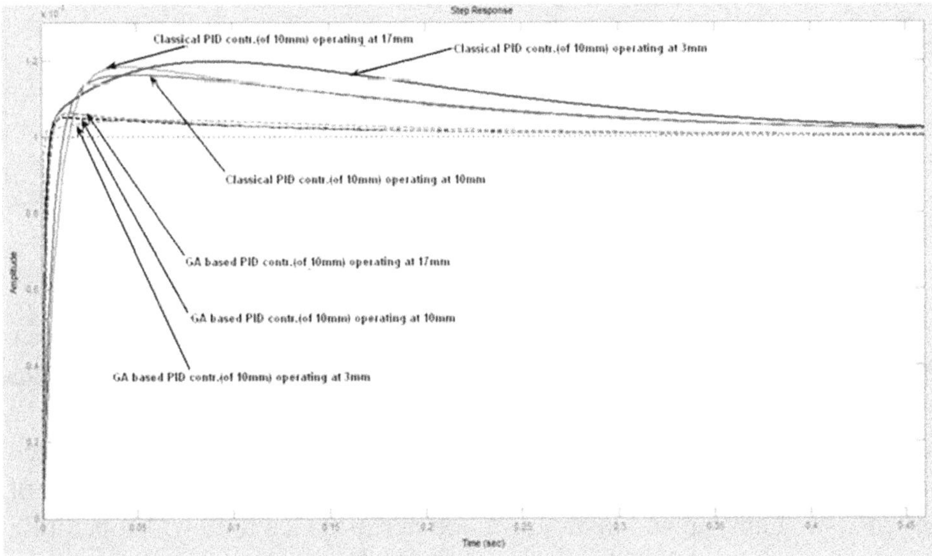

Table 5. Comparative study between dynamic performances with classical and GA based nominal PID controller operating at nominal and other off-nominal (low and high) air-gaps

Air Gap	Specifications	Classical	GA
	Peak Overshoot(%)	19.7	4.3
	Settling Time(sec)	0.471	0.207
3 mm (low)	Rise Time(sec)	0.003	0.00116
	S.S.Error	0	0
	Peak Overshoot(%)	16.1	4.96
	Settling Time(sec)	0.433	0.147
10 mm (nominal)	Rise Time(sec)	0.00813	0.00299
	S.S.Error	0	0
	Peak Overshoot(%)	18.3	6.11
	Settling Time(sec)	0.427	0.144
17 mm (high)	Rise Time(sec)	0.00992	0.00372
	S.S. Error	0	0

Table 6. Comparative parameters of PID Controller at 10 mm air-gap

Parameters	Classical	GA
Gain (K)	0.0145	0.04688
1st Zero (Z1)	21	25.8638
2nd Zero (Z2)	5	6.9794
P-gain	0.377	1.54
I-gain	1.5225	8.4639
D-gain	0.0145	0.0469

From the simulated results, it is found that the classical controllers provide satisfactory performance at desired operating point (nominal), but, performance degradation occurs as the operating point (off-nominal) is changed. It becomes difficult for the controller to maintain satisfactory performances if there is wide variation of operating air-gap. On the other hand, GA based optimized controllers shows better performance than classical controllers irrespective of any operating air-gap and maintains good performance for a wide operating area (3 mm to 17 mm) of the proposed DCALS.

Among optimized controllers, the PID controller shows best performance in nominal as well as off-nominal operating air-gap. It not only reduces overshoots but also shows improved speed of response (less rise time) and produces zero steady-state error irrespective of any operating air-gap.

Figure 8 shows the comparison between performances of classical and optimized controllers at nominal operating air-gap (10 mm). It is clear that the GA based optimized controllers (Lead, Lag-Lead and PID) shows better performance than corresponding three classically designed controllers.

Convergence of the PID parameters such as gain and the two zeros has been shown here in Figure 9 to Figure 11. It is seen (Figure 12) that the objective function minimization process is becoming almost constant after 30th iterations since the chromosomes in the population of GA is having similar structure and having high value of fitness, so that, no amount of gene mixing is able to produce better result. This is a drawback of GA and hence, the results are not fully optimal but sub-optimal.

Figure 8. Comparative position response of DCALS with classical and GA based controllers (Lead, Lead-Lag, PID) at 10 mm operating air-gap

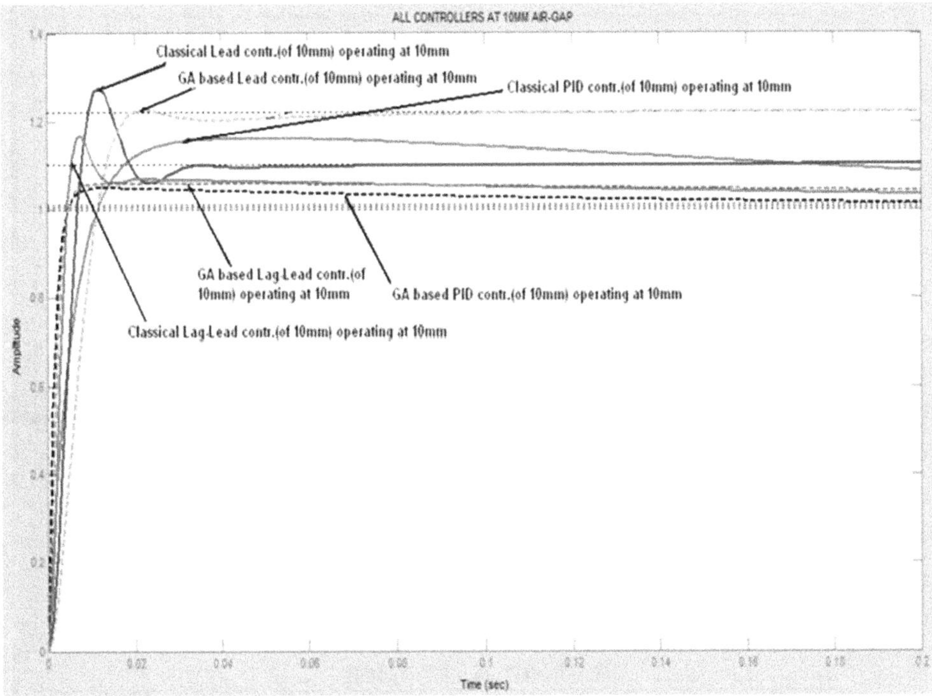

Figure 9. PID controller's gain K vs iteration

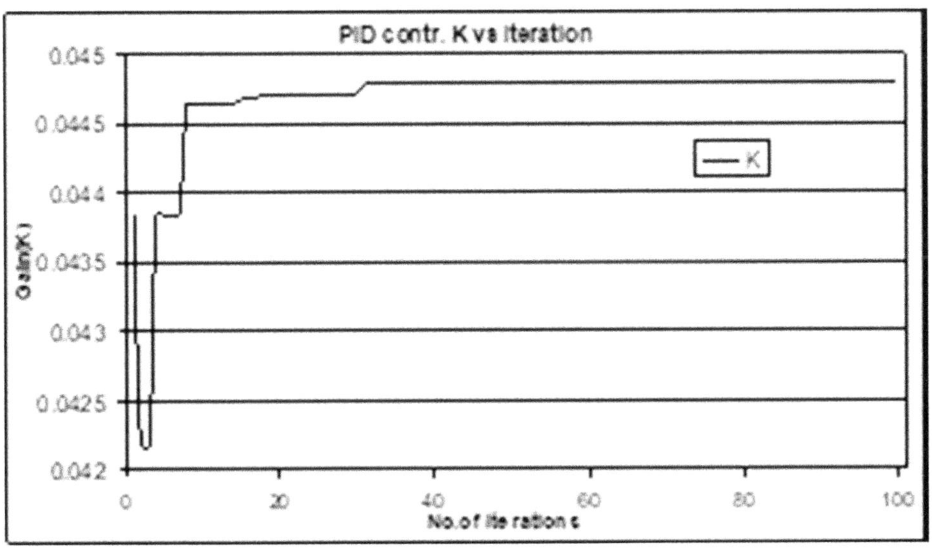

Figure 10. PID controller's zero Z1 vs iteration

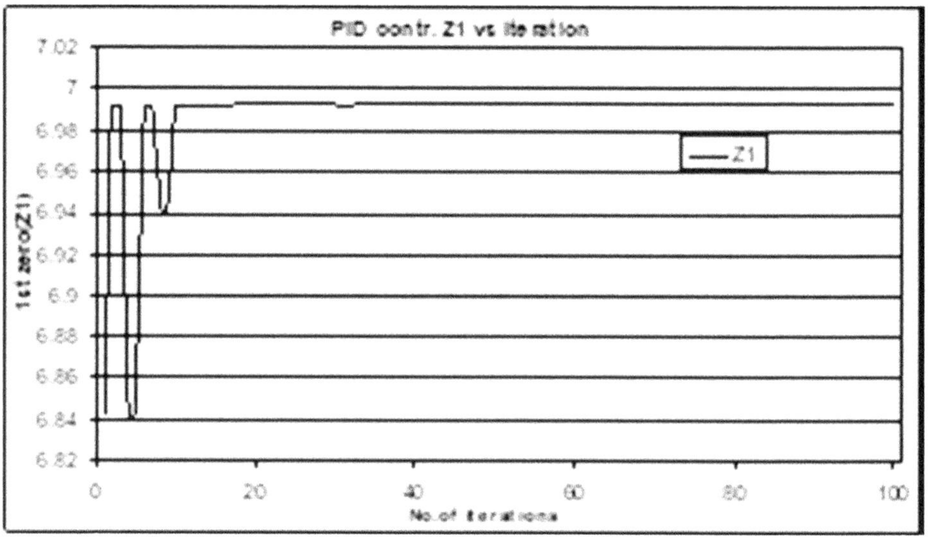

Figure 11. PID controller's zero Z2 vs iteration

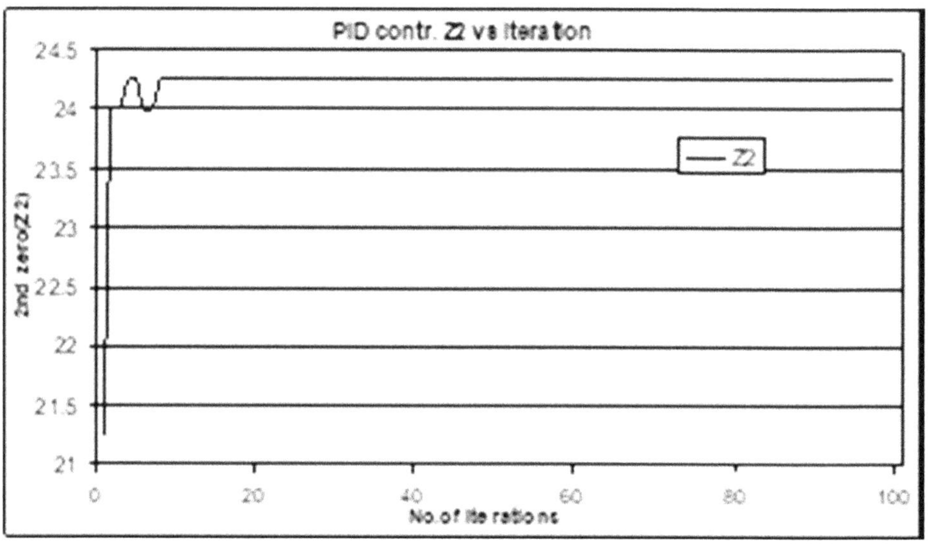

Figure 12. Objective function vs. no of iterations for GA based PID controller

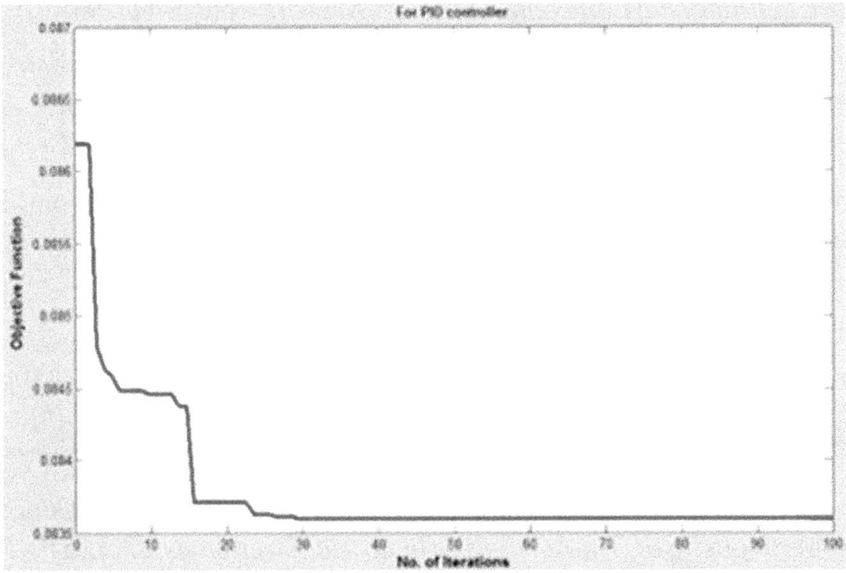

REFERENCES

Cadence Design Systems. (2000). *Orcad capture user's guide* (2nd ed.). Author.

Gaing, Z.-L. (2004). A particle swarm optimization approach for optimum design of PID controller in AVR system. *IEEE Transactions on Energy Conversion*, *19*(2), 384–391. doi:10.1109/TEC.2003.821821

Goldberg, D. E. (1989). *Genetic algorithms in search, optimization, and machine learning*. Addison-Wesley Longman Publishing Co., Inc.

Gundogdu, O. (2005). Optimal-tuning of PID controller gains using genetic algorithms. *Journal of Engineering Sciences*, *11*(1), 131–135.

Hu, X. (2006). *Particle swarm optimization*. Tutorial of the IEEE Swarm Intelligence Symposium.

Jan, J. A., & Sulc, B. (2002, May 27–30). Evolutionary computing methods for optimizing virtual reality process models. In *International Carpathian Control. Conference (ICCC 2002)*. Malenovice.

Joo, S., & Seo, J. H. (1997). Design and analysis of the nonlinear feedback linearizing control for an electromagnetic suspension system. *IEEE Transactions on Control Systems Technology*, *5*(1), 593–598.

Kennedy, J., & Eberhart, R. (1995, November). Particle swarm optimization. In *IEEE International Conference on Neural Networks* (pp. 1942–1948). IEEE.

Kim, D. H., & Park, J. I., III. (2005). Intelligent PID controller tuning of AVR system using GA and PSO. Lecture Notes in Computer Science, 375. doi:10.1007/11538356_38

Krohling, R. A., Jaschek, H., & Rey, J. P. (1997). Designing PI/PID controller for a motion control system based on genetic algorithm. In *Proceedings of the 12th IEEE International Symposium on Intelligent Control* (pp. 125–130). 10.1109/ISIC.1997.626429

Mahony, T. O., Downing, C. J., & Fatla, K. (2000). Genetic algorithm for PID parameter optimization: Minimizing error criteria. *Process Control. and Instrumentation*, 148–153.

Man, K. F., Tang, K. S., & Kwong, S. (1996). Genetic algorithms: Concepts and applications. *IEEE Transactions on Industrial Electronics*, *43*(5), 519–534. doi:10.1109/41.538609

Rathikarani, D., Sivakumar, D., & Mary, S. A. J. (2007). Genetic Algorithm based Optimization of Controller Tuning for Processes with Dead Time. Advances in Control and Optimization of Dynamical Systems.

Sadasivarao, M. V., & Chidambaram, M. (2006). PID controller tuning of cascade control systems using genetic algorithm. *Journal of the Indian Institute of Science*, 343–354.

Salami, M., & Cain, G. (1995). An adaptive PID controller based on genetic algorithm processor. Genetic Algorithms in Engineering Systems, (414). doi:10.1049/cp:19951030

Shin, G. W., Song, Y. J., Lee, T. B., & Choi, H. K. (2007). Genetic algorithm for identification of time delay systems from step responses. *International Journal of Control, Automation, and Systems*, *5*(1), 79–85.

Sinha, P. K. (1987). *Electromagnetic suspension, dynamics and control.* Peter Peregrinus Ltd.

Trumper, D. L., Olson, S. M., & Subrahmanyan, P. K. (1997). Linearizing control of magnetic suspension systems. *IEEE Transactions on Control Systems Technology*, *5*(4), 427–438. doi:10.1109/87.595924

Chapter 10
Optimization of Controller Parameters for the Proposed DCALS Using Particle Swarm Optimization (PSO) Technique

ABSTRACT

GA is a well-established algorithm with many versions and many applications. PSO is a relatively recent heuristic search method whose mechanics are inspired by the swarming or collaborative behavior of biological populations. PSO is similar to the GA in the sense that these two evolutionary heuristics are population-based search methods. In other words, PSO and the GA move from a set of points (population) to another set of points in a single iteration with likely improvement using a combination of deterministic and probabilistic rules.

PARTICLE SWARM OPTIMISATION (PSO) OVERVIEW

While the GA can rapidly locate good solutions, even for difficult search spaces, it has some disadvantages associated with it: 1) unless the fitness function is defined properly, GA may have a tendency to converge towards local optima rather than the global optimum of the problem; 2) operating on dynamic data sets are difficult and 3) for specific optimization problems, and given the same amount of computation time, simpler optimization algorithms

DOI: 10.4018/978-1-6684-7388-7.ch010

may find better solutions than Gas (del Valle, et. al. 2008). Deficiencies in GA performance are inevitable since, the crossover and mutation operations cannot ensure better fitness of offspring because chromosomes in the population have similar structure and their average fitness are high toward the end of the evolutionary process (Fogel, 1995; Eberhart et. al. 1998; Nasri, et. al. 2007).

PSO is an algorithm modeled on swarm intelligence that finds a solution to an optimization problem in a search space or model and predicts social behavior in the presence of objectives. The PSO is a stochastic, population-based computer algorithm modeled on swarm intelligence. Swarm intelligence is based on social-psychological principles and provides insights into social behavior, as well as contributing to engineering applications. (Kennedy et. al. 1995; Eberhart, et. al. 1995; Hu, 2006).

PSO can converge to the near optimal solution in many problems where most analytical methods fail to converge. This technique was introduced and developed by Dr. Eberhart and Dr. Kennedy in 1995, inspired by social behavior of bird flocking (Figure 1) or, fish schooling (Figure 2) (Kennedy, et. al. 1995; Hu, 2006; Abido, 2002). Generally, PSO is characterized as a simple concept, easy to implement, and computationally efficient. Unlike the other heuristic techniques, PSO has a flexible and well-balanced mechanism to enhance the global and local exploration abilities (Shi, et. al. 1998).

PSO has some advantages over other similar optimization techniques such as GA, which are as follows (del Valle, et. al. 2008; Eberhart, et. al. 1998):

1) PSO is easier to implement and there are fewer parameters to adjust.
2) In PSO, every particle remembers its own previous best value as well as the neighborhood best; therefore, it has a more effective memory capability than GA.
3) PSO is more efficient in maintaining the diversity of the swarm (Abido, 2002; Yoshida, et. al. 2000) (more similar to the ideal social interaction in a community), since all the particles use the information related to the most successful particle in order to improve themselves, whereas in GA, the worse solutions are discarded and only the good ones are saved; therefore, in GA the population evolves around a subset of the best individuals. Compared with GA, all the particles in PSO tend to converge to the best solution quickly, even in the local version in most cases.

Figure 1. Photograph of bird flocking

Figure 2. Photograph of fish schooling

The PSO method has been found to be robust in solving problems featuring nonlinearity and non-differentiability, multiple optima, and high dimensionality through adaptation, which is derived from the social-psychological theory (Gundogdu, 2005; Shi, et. al. 1999). According to the research results for a flock of birds, birds find food by flocking (not by each individual). The observation leads to the assumption that every information is shared inside flocking. Moreover, according to observation of behavior of human groups, behavior of each individual (agent) is also based on behavior patterns authorized by the groups such as customs and other behavior patterns according to the experiences by each individual (Engelbrecht, 2006; Shi, et. al. 1998). In the PSO algorithm, instead of using evolutionary operators such as mutation and crossover, to manipulate algorithms, for a d-variabled optimization problem, a flock of particles are put into the d-dimensional search space with randomly chosen velocities and positions knowing their best values so far (pbest) and the position in the d-dimensional space (Shi, et. al. 1998). The velocity of each particle is adjusted according to its own flying experience and the other particles flying experience.

The PSO algorithm is population-based: a set of potential solutions evolves to approach a convenient solution (or set of solutions) for a problem. Being an optimization method, the aim is finding the global optimum of a real-valued function (fitness function) defined in a given space (search space).

The social metaphor that has led to this algorithm can be summarized as follows: the individuals that are part of a society hold an opinion that is part of a "belief space" (the search space) shared by every possible individual. Individuals may modify this "opinion state" based on some factors such as:

- The knowledge of the environment (its fitness value)
- The individual's previous history of states (its memory)
- The previous history of states of the individuals themselves (pbest)
- The previous history of states of the individual's neighborhood (gbest)

An individual's neighborhood may be defined in several ways, configuring somehow the "social network" of the individual. Several neighborhood topologies exist (full, ring, star, etc.) depending on whether an individual interacts with all, some, or only one of the rest of the population.

Following certain rules of interaction, the individuals in the population adapt their scheme of belief to the ones that are more successful among their social network. Over the time, a culture arises, in which the individuals hold opinions that are closely related.

OPTIMIZATION OF CONTROLLER PARAMETERS USING PSO

In the PSO algorithm each individual is called a "particle", and is subject to a movement in a multidimensional space that represents the belief space. Particles have memory, thus retaining part of their previous state. There is no restriction for particles to share the same point in belief space, but in any case their individuality is preserved. Each particle's movement is the composition of a previous velocity and two randomly weighted influences: individuality, the tendency to return to the particle's best previous position, and sociality, the tendency to move towards the neighborhood's best previous position. The flow chart for the PSO program has been provided in the Figure 3.

Figure 3. Flow chart for the PSO program

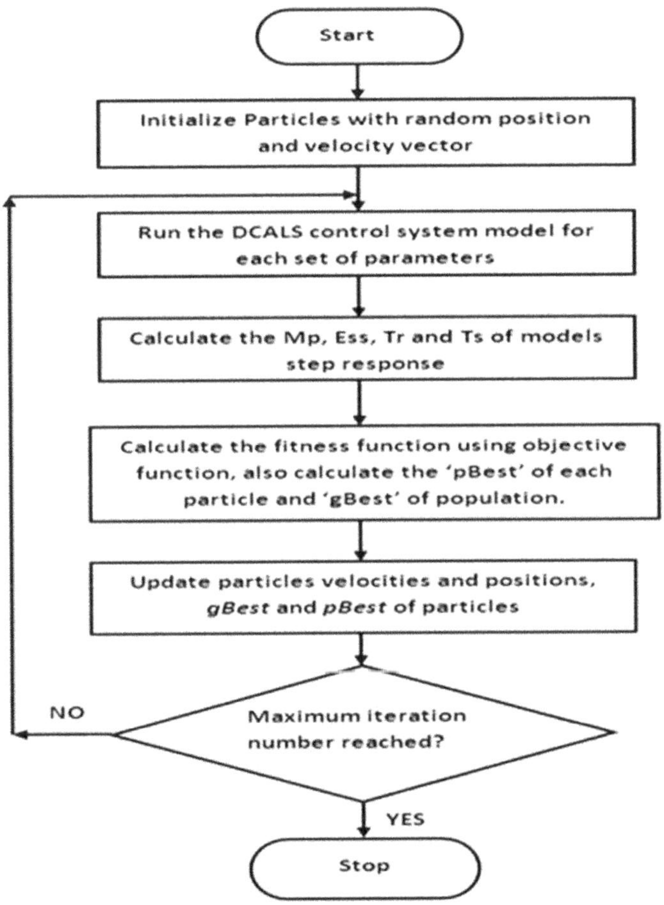

FORMULATION TECHNIQUE AND PSO IMPLEMENTATION

In this part of the work the modified form of PSO algorithm has been implemented (Yoshida, 2000) for better convergence. For proper convergence purpose the acceleration coefficients (constants) should be set sufficiently high, but higher acceleration coefficients result in less stable systems in which the velocity has a tendency to explode, therefore, to fix this problem, the velocity v_i is usually kept within the range [*-vmax*, *+vmax*], but still neither limiting the velocity necessarily prevents particles from leaving the search space, nor it helps to guarantee convergence. An inertia weight 'w' is introduced to control the velocity explosion. Now, if w, c_1 and c_2 are set correctly, this update rule allows for convergence without the use of *vmax*. The inertia weight can be used to control the balance between exploration and exploitation (Yoshida, 2000):

$w \geq 1$: velocities increase over time, swarm diverges
$0 < w < 1$: particles decelerate, convergence depends on c_1 and c_2.

Essentially, this parameter '*w*' controls the exploration of the search space, therefore an initially higher value (typically 0.9) allows the particles to move freely in order to find the global optimum neighborhood fast. Once the optimal region is found, the value of the inertia weight can be decreased (usually to 0.4) in order to narrow the search, shifting from an exploratory mode to an exploitative mode. Commonly, a linearly decreasing inertia weight (first introduced by Shi and Eberhart, 1998 & 1999) has produced good results in many applications; however, the main disadvantage of this method is that once the inertia weight is decreased, the swarm loses its ability to search new areas because it is not able to recover its exploration mode.

In the present approach the constant inertia weight factor has been implemented and, the velocity of each particle is adjusted according to its own flying experience and the other particles flying experiences. For example, the *i* th particle is represented as $x_i = (x_{i,1}, x_{i,2}, \ldots, x_{i,d})$ in the d-dimensional space. The best previous position of the *i* th particle is recorded and represented as: $Pbest_i = (Pbest_{i,1}, Pbest_{i,2}, \ldots, Pbest_{i,d})$

The index of the best particle among all the particles in the group is $gbest_d$. The velocity for particle *i* is represented as $v_i = (v_{i,1}, v_{i,2}, \ldots, v_{i,d})$. The modified velocity and position of each particle can be calculated using the current velocity and the distance from $Pbest_{i,d}$ to $gbest_d$ as shown in the following formulae (Nasri, et. al. 2007; Gundogdu, 2005, Yoshida, 2000):

$$v_{i,m}^{(t+1)} = w.v_{i,m}^{(t)} + c_1 * rand_1(\ ) * (pbest_{i,m} - x_{i,m}^{(t)}) + c_2 * rand_2(\ ) * (gbest_m - x_{i,m}^{(t)}) \quad (9.18)$$

$$x_{i,m}^{(t+1)} = x_{i,m}^{(t)} + v_{i,m}^{(t+1)} \quad (9.19)$$

for, i = 1,2,….n, and m = 1,2,….,d.

where, n Number of particles
d Dimension
t Pointer of iterations (generations)
$v_{i,m}^{(t)}$ Velocity of particle i at t^{th} iteration, $V_d^{\min} \leq v_{i,d}^{(t)} \leq V_d^{\max}$
w Inertia weight factor
c_1, c_2 Acceleration constant / coefficients
rand $_1$(), Random number between 0 and 1
rand $_2$()
$x_{i,m}^{(t)}$ Current position of particle i at t^{th} iteration
$pbest_i$ Best previous position of the ith particle
gbest Best particle among all the particles in the population

OBJECTIVE FUNCTION IMPLEMENTATION FOR PSO

In Section 10.3.1 different integral performance criterions have been mentioned which have their own advantages and disadvantages. For example, disadvantage of the IAE and ISE criteria is that their minimizations can result in responses with relatively small overshoot but with a long settling time because the said performance criterions weight all errors equally independent of time. Although both the ITAE and ITSE performance criterion can overcome the disadvantage of the IAE and ISE criterion, the derivation processes of the analytical formula are complex and time-consuming (Nasri, et. al. 2007; Krohling, 1997).

In this part of the work while implementing PSO a time domain criterion has been utilized as the objective function. A proper range of parameters of the designed controllers yield a good step response. This will result in performance criteria minimization in the time domain, where the performance criteria in the time domain include overshoot, rise time, settling time and steady state error.

The objective function is defined by Eqn.10.20,

$$\min_{K:\text{stabilizing}} W(K) = (1 - e^{-\beta}).(M_P + E_{SS}) + e^{-\beta}.(t_S - t_r) \tag{10.20}$$

where, K is the controller gain which is d-dimensional matrix and β is the weighting factor. M_p is the percentage overshoot, E_{SS} is the steady-state error, t_S is the settling time, t_r is the rise time.

Normally, the desired performance can be achieved out of this performance criterion W(K) by choosing a proper value of the weighting factor β. The optimum selection of β depends on the designer's requirement and the characteristics of the plant under control. To reduce overshoot and steady state errors, β could be set higher than 0.7 and to reduce rise time and settling time it is to be set lower than 0.7; for the present work β is taken as 0.8 (Nasri, 2007; Gundogdu, 2005).

TERMINATION CRITERION FOR PSO PROGRAM

Here, the termination criteria have been considered to be the attainment of the maximum number of iterations at 100.

RESULTS OBTAINED FROM PSO OPTIMIZATION

The PSO has been implemented with the similar philosophy of design as that was in case of GA (Bhaduri, Bagchi, & Banerjee, 2017). The structure of the resulting controller is almost identical to that was obtained from GA. Only the controller parameters are changed. It is also expected to improve transient response since optimized parameters of the controllers occur more closely to the global optima of the entire search space in the possible solution area.

The parameters used for this optimization are as follows:

- Population size: 50
- Iterations: 100
- Inertia weight factor, w = 0.6;
- Acceleration constants, c_1, c_2 = 1.70;

A comparative position responses of the proposed DCALS with classical and PSO based nominal Lead controller operating at nominal and other off-nominal (low and high) air-gaps have shown in Figure 4. It is seen that PSO based optimized nominal controllers have been demonstrated better transient performance than corresponding classical Lead controllers over a large variation of air-gap (3mm to 17 mm). It is to be mentioned that Lead controller cannot reduce steady-state error. Again percent overshoot and steady-state error are the two conflicting parameters, so while doing optimization of controller parameters it is aimed to improve transient performance sacrificing little bit steady-state accuracy.

Figure 4. Comparative position responses of DCALS with classical and PSO based nominal Lead controller operating at nominal and off-nominal (low and high) air-gaps

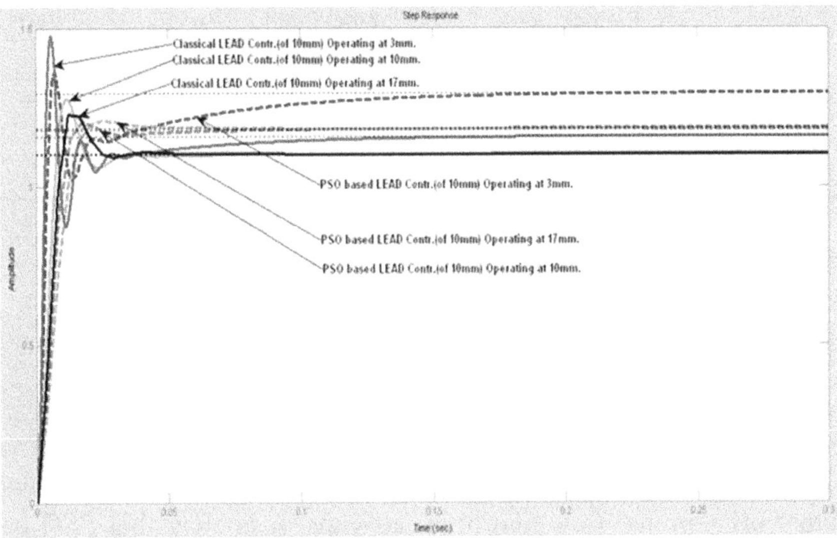

Table 1 shows a comparative position response of DCALS with classical, GA and PSO based nominal Lead controllers while operating at nominal and other off-nominal (low and high) air-gaps. It is obvious that the PSO algorithm is more competent in finding optimal results than GA. PSO based Lead controller reduces overshoot with faster transient performance (less rise time). Due to obvious reasons (as mentioned), steady-state error are relatively more in position response with optimized controllers (GA and PSO) than the response obtained with classical conventional controllers. Table 2 shows the comparative parameters of Lead Controller at 10 mm gap.

Table 1. Comparative performances with classical, GA and PSO based nominal Lead controller operating at nominal and off-nominal (low and high) air-gaps

Air Gap	Specifications	Classical	GA	PSO
	Peak Overshoot(%)	29.5	9.6	5.2
	Settling Time(sec)	0.0767	0.0527	0.114
3mm	Rise Time(sec)	0.00252	0.00563	0.00386
	S.S.Error	0.16	0.37	0.29
	Peak Overshoot(%)	16	0.443	1.85
	Settling Time(sec)	0.0281	0.0172	0.0138
10mm (nominal)	Rise Time(sec)	0.00536	0.0111	0.0104
	S.S.Error	0.1	0.22	0.18
	Peak Overshoot(%)	13.2	2.26	2.51
	Settling Time(sec)	0.0229	0.0381	0.0306
17mm	Rise Time(sec)	0.00653	0.014	0.0116
	S.S.Error	0.1	0.22	0.18

Table 2. Comparative parameters of lead controller at 10 mm gap

Parameters	Classical	GA	PSO
Gain (K)	7.3	4.0109	5.202
Zero (Z)	25	24.9409	24.7062
Pole (P)	323	349.422	380.5204

Figure 5 shows comparative position responses of the proposed DCALS with classical and PSO based nominal PID controller operating at nominal and other off-nominal (low and high) air-gaps. It is clear that PSO based optimized nominal PID controller (designed for 10 mm air-gap) has shown better performance than corresponding conventional classical PID controller irrespective of any operating air-gap.

Table 3 shows a comparative position response of DCALS with classical, GA and PSO based nominal PID controllers while operating at nominal (10 mm) and off-nominal (low and high) air-gaps. It is seen that PSO based optimized nominal PID controller shows excellent performance (both in transient and steady-state) over a large operating air-gap (3 mm to 17 mm). The comparative parameters of PID Controller at 10 mm gap are shown in Table 4.

Figure 5. Comparative position responses of DCALS with classical and PSO based nominal PID controller operating at nominal and off-nominal (low and high) air-gaps

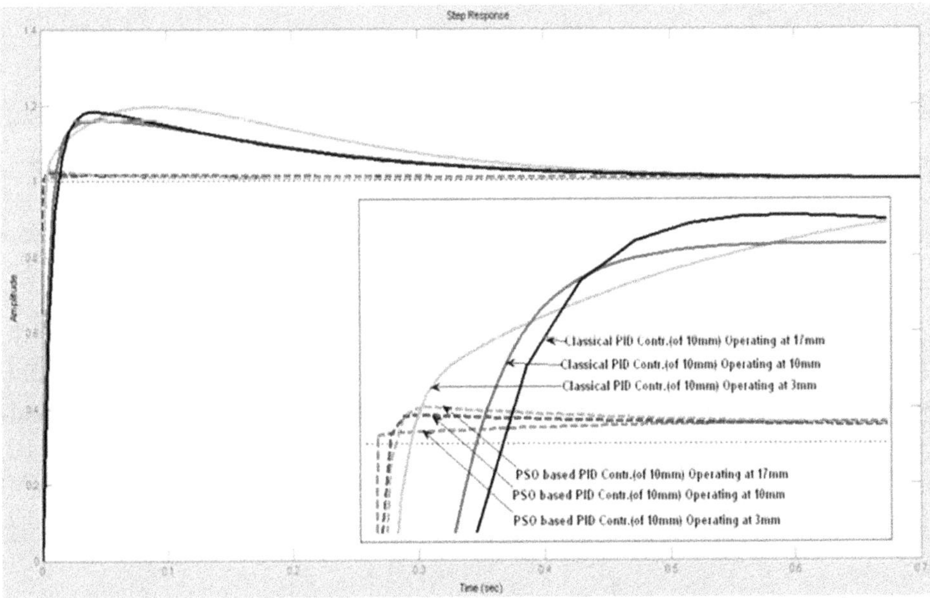

Table 3. Comparative performances with classical, GA and PSO based nominal PID controller operating at nominal and other off-nominal (low and high) air-gaps

Air Gap	Specifications	Classical	GA	PSO
	Peak Overshoot(%)	19.7	4.3	1.76
	Settling Time(sec)	0.471	0.207	0.000862
3mm	Rise Time(sec)	0.003	0.00116	0.000552
(low)	S.S.Error	0	0	0
	Peak Overshoot(%)	16.1	4.96	2.34
	Settling Time(sec)	0.433	0.147	0.0188
10mm	Rise Time(sec)	0.00813	0.00299	0.00135
(nominal)	S.S.Error	0	0	0
	Peak Overshoot(%)	18.3	6.11	2.94
	Settling Time(sec)	0.427	0.144	0.0322
17mm	Rise Time(sec)	0.00992	0.00372	0.00177
(high)	S.S. Error	0	0	0

Table 4. Comparative parameters of PID Controller at 10 mm gap

PID Parameter	Classical	GA	PSO
Gain (K)	0.0145	0.0469	0.1076
1st Zero (Z1)	21	25.8638	34.5613
2nd Zero (Z2)	5	6.9794	2.4817
P-gain	0.377	1.54	3.9858
I-gain	1.5225	8.4639	9.2303
D-gain	0.0145	0.0469	0.1076

Similarly, a comparative position responses of the proposed DCALS with classical and PSO based nominal Lead-Lag controller operating at nominal and other off-nominal (low and high) air-gaps is shown in Figure 6. It is clear that PSO based optimized nominal Lead-Lag controller (designed for 10 mm air-gap) has shown better performance than corresponding conventional classical Lead-Lag controller irrespective of any operating air-gap.

Figure 6. Comparative position responses of DCALS with classical and PSO based nominal Lead-Lag controller operating at nominal and off-nominal (low and high) air-gaps

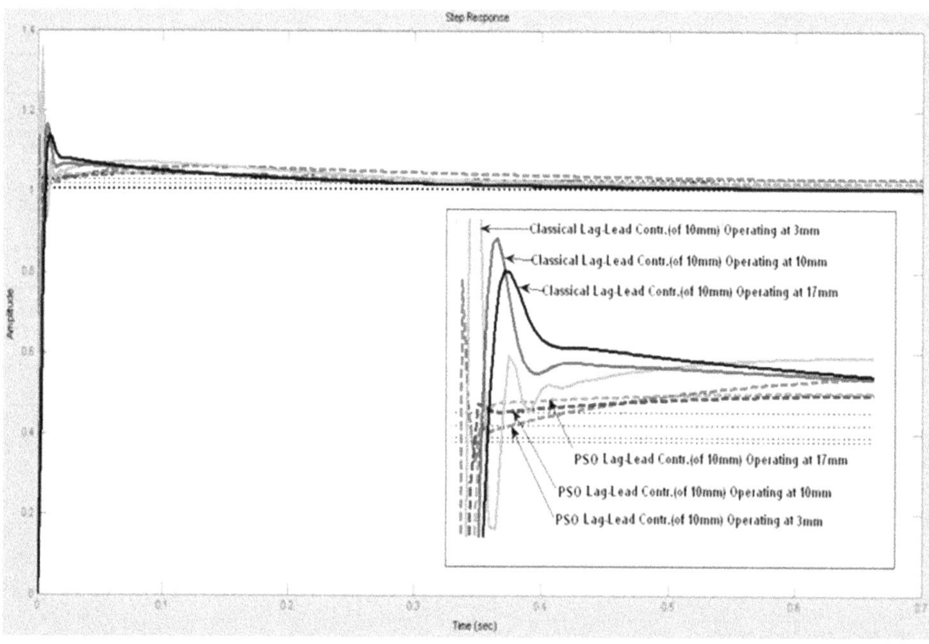

A comparative position response of DCALS with classical, GA and PSO based nominal Lead-Lag controllers while operating at nominal (10 mm) and off-nominal (low and high) air-gaps. From the results of Table 5, it is clear that PSO based optimized Lead-Lag controllers shows better transient performance (not in steady-state) than corresponding classical and GA based controllers. It is to be mentioned that with the use of Lead-Lag controller steady-state error cannot be reduced to zero. Table 6 shows comparative parameters of Lead-Lag Controller at 10 mm gap

Table 5. Comparative performances with classical, GA and PSO based nominal Lead-Lag controller operating at nominal and other off-nominal (low and high) air-gaps

Air Gaps	Specifications	Classical	GA	PSO
	Peak Overshoot(%)	34.4	8.97	8.0
	Settling Time(sec)	0.332	0.443	0.284
3mm	Rise Time(sec)	0.0017	0.00124	0.00087
	S.S.Error	0.01	0.01	0.03
	Peak Overshoot(%)	16.1	4.65	2.39
	Settling Time(sec)	0.254	0.338	0.181
10mm	Rise Time(sec)	0.00333	0.00304	0.00221
(nominal)	S.S.Error	0.01	0.01	0.02
	Peak Overshoot(%)	13.5	5.84	2.54
	Settling Time(sec)	0.248	0.331	0.179
17mm	Rise Time(sec)	0.0041	0.0040	0.00301
	S.S.Error	0.01	0.01	0.02

Table 6. Comparative parameters of lead-lag controller at 10mm gap

Parameters	Classical	GA	PSO
Gain (K)	21.2	68.949	114.9
Lead Zero (Z1)	25	24.3039	17.1
Lead Pole (P1)	665	1922.1	2093.7
Lag Zero(Z2)	5	3.2397	4.6
Lag Pole(P2)	0.5	0.5611	1.6

A comparison between the position responses of DCALS at nominal operating air-gap with conventional classical, GA and PSO based Lead, Lead-Lag and PID controller are shown in Figure 7, Figure 8 and Figure 9 respectively. It appears that PSO based PID controller shows the best performance at the desired operating air-gap. The similar comparative results already been tabulated in Tables 1, 3 and 5 respectively.

Figure 10 displays the convergence of the objective function with respect to iterations in case of PSO based PID controller, which depicts that PSO convergence is much faster than that of GA (Figure 12 in Chapter 9). Here convergence towards optimal solution occurs after 20th iterations.

This is the virtue of PSO based algorithms since; they always have effective memory capability than the chromosomes in GA. All the particles in PSO use the information related to the most successful particle in order to improve and adjust themselves and thereby each particle is stochastically accelerated toward the global best position (of the entire population) or, to the neighborhoods (local) best position (Bhaduri, Bagchi, & Banerjee, 2017).

Figure 7. Comparative position responses of DCALS with classical, GA and PSO based Lead controllers at 10 mm air-gap

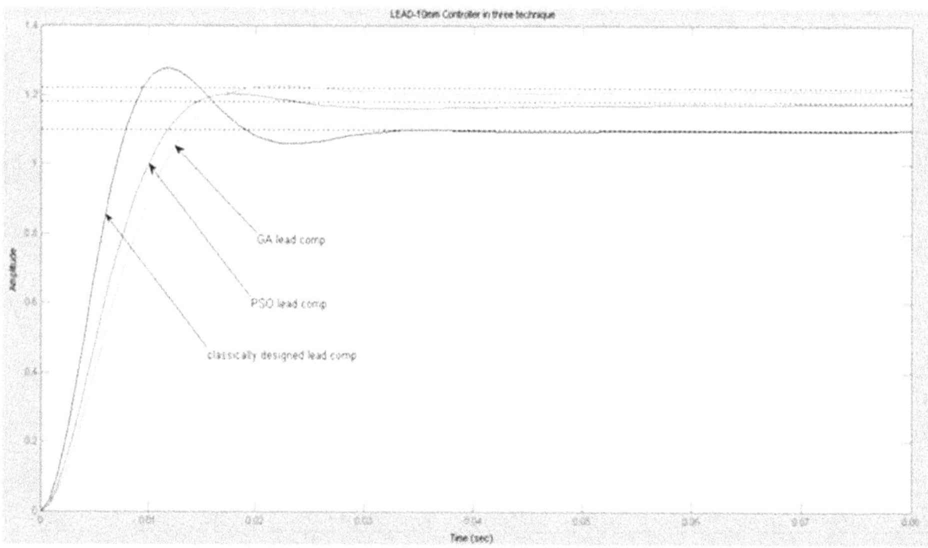

Figure 8. Comparative position responses of DCALS with classical, GA and PSO based Lead-Lag controllers at 10 mm air-gap

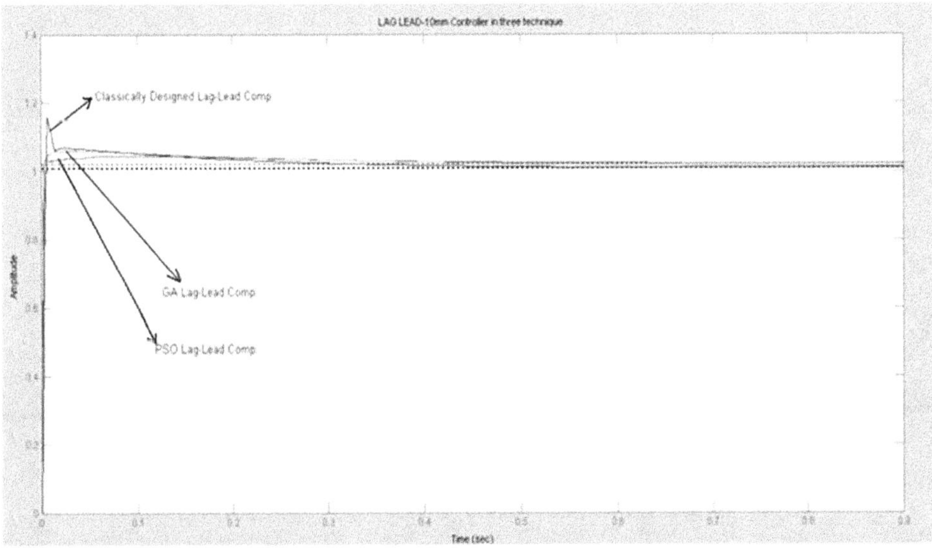

Figure 9. Comparative position responses of DCALS with classical, GA and PSO based PID controllers at 10 mm air-gap

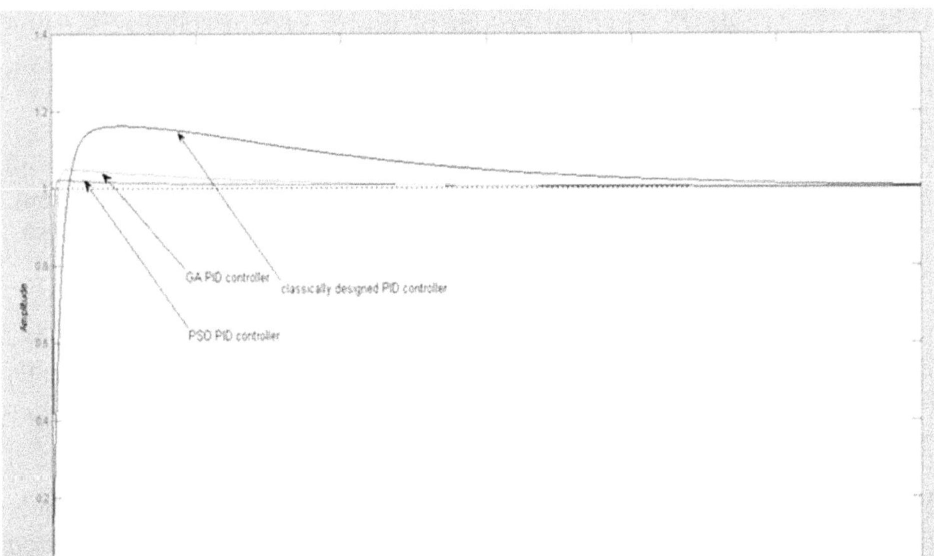

Figure 10. Objective function vs. no of iterations for PSO based PID controller

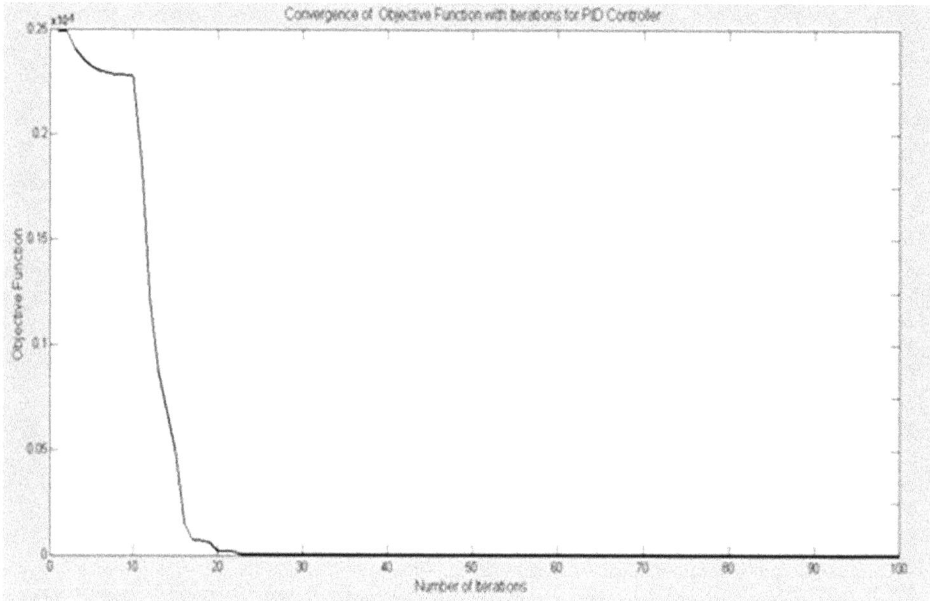

REFERENCES

Abido, M. A. (2002). Optimal design of power-system stabilizers using particle swarm optimization. *IEEE Transactions on Energy Conversion*, *17*(3), 406–413. doi:10.1109/TEC.2002.801992

Bhaduri, R., Bagchi, S., & Banerjee, S. (2017, April-May). PSO Optimization of the Controller Parameters for an Inherently Magnetic Levitation System. *IACSIT International Journal of Engineering and Technology*, *9*(2), 2319–8613.

del Valle, Y. D., Venayagamoorthy, G. K., Mohagheghi, S., Hernandez, J.-C., & Harley, R. G. (2008). Particle swarm optimization: Basic concepts, variants and applications in power systems. *IEEE Transactions on Evolutionary Computation*, *12*(2), 171–195. doi:10.1109/TEVC.2007.896686

Eberhart, R., & Kennedy, J. (1995, October). A new optimizer using particle swarm theory. In *6th International Symposium on Micro Machine and Human Science* (pp. 39–43). 10.1109/MHS.1995.494215

Eberhart, R. C., & Shi, Y. (1998, May). Comparison between genetic algorithms and particle swarm optimization. In *IEEE International Conference on Evolutionary Computation* (pp. 611–616). IEEE. 10.1007/BFb0040812

Engelbrecht, A. P. (2006). Particle swarm optimization: Where does it belong? In *Proceedings of the IEEE Swarm Intelligence Symposium* (pp. 48–54). IEEE.

Fogel, D. B. (1995). *Evolutionary computation toward a new philosophy of machine intelligence*. IEEE Publications.

Gundogdu, O. (2005). Optimal-tuning of PID controller gains using genetic algorithms. *Journal of Engineering Sciences*, *11*(1), 131–135.

Hu, X. (2006). *Particle swarm optimization*. Tutorial of the IEEE Swarm Intelligence Symposium.

Kennedy, J., & Eberhart, R. (1995, November). Particle swarm optimization. In *IEEE International Conference on Neural Networks* (pp. 1942–1948). IEEE.

Krohling, R. A., Jaschek, H., & Rey, J. P. (1997). Designing PI/PID controller for a motion control system based on genetic algorithm. In *Proceedings of the 12th IEEE International Symposium on Intelligent Control* (pp. 125–130). 10.1109/ISIC.1997.626429

Nasri, M., Nezamabadi-pour, H., & Maghfoori, M. (2007). A PSO-based optimum design of PID controller for a linear brushless DC motor. *World Academy of Science, Engineering and Technology*, 211–215.

Shi, Y., & Eberhart, R. (1998a). A modified particle swarm optimizer. In *Proceedings of the IEEE International Conference on Evolutionary Computation* (pp. 69–73). IEEE Publications.

Shi, Y., & Eberhart, R. (1998b). Parameter selection in particle swarm optimization. In *Proceedings of the 7th International Conference on Evolutionary Programming (EP)* (pp. 591–600). IEEE.

Shi, Y., & Eberhart, R. C. (1999c). Empirical study of particle swarm optimization. In *Proceedings of the IEEE Congress on Evolutionary Computation* (pp. 1945–1950). IEEE. 10.1109/CEC.1999.785511

Yoshida, H., Kawata, K., Fukuyama, Y., Takayama, S., & Nakanishi, Y. (2000). A particle swarm optimization for reactive power and voltage control considering voltage security assessment. *IEEE Transactions on Power Systems*, *15*(4), 1232–1239. doi:10.1109/59.898095

Chapter 11
Conclusions and Scope for Further Work

ABSTRACT

As the name suggests, the work of this book is based on some studies of DC attraction type levitation (DCALS) and associated control systems. For the successful implementation of any DCALS, one should require the use of fundamentally interdisciplinary concepts of electromagnetism, electronics, power electronics, mechanical engineering, measurement, and control systems. Design, implementation, and testing of a DC attraction type levitation system where a ferromagnetic cylindrical rod of around 0.122 kg mass is made to remain suspended at the desired operating gap under two I-core electromagnets that have been controlled independently through two identical controllers and the stable levitation of the rod is achieved through SISO control of each air-gap corner.

CONCLUSION

As the name suggests, the work of this book is based on some studies of DC attraction type levitation (DCALS) and associated control systems. For the successful implementation of any DCALS, one should require the use of fundamentally interdisciplinary concepts of electromagnetism, electronics, power electronics, mechanical engineering, measurement and control systems. DCALS primarily consists of four major components: such as (i) electromagnets (ii) power amplifiers, (iii) controllers and (iv) sensors. But

DOI: 10.4018/978-1-6684-7388-7.ch011

design and implementation of power amplifiers and controllers plays most important part of work for any maglev project. In this book the emphasis is given on design and implementation of power amplifier and controller for the proposed DCALS. This book deals with the design and development of electronic controllers for stable levitation of DCALS. Description of the experimental setup, modeling, design of controller, hardware circuits and power amplifiers have been presented. The controller design has been validated by simulation and practical implementation.

A simplified analytical model of the DCALS has been derived from the basic dynamic equations. The DCALS is highly non-linear in nature due to the magnet force-distance-current characteristics. After making appropriate assumptions, the non-linear equations are linearised about an operating point and a linear model has been derived for the proposed system. After estimating the system parameters the transfer functions of the system at different operating points have been determined. Since DCALS is an inherently unstable system, the stabilization of the overall closed loop system is the main objective of this work. A cascade compensation control scheme utilizing an inner current loop and an outer position loop has been developed. The simulated controllers have been successfully implemented, thus validating the modelling and controller design procedure.

Power amplifier is one of the important components for the implementation of any DCALS. DCALS critically deserves fast response and the switch mode power amplifiers used for such applications have important role to meet this demand. Different possible switched mode power amplifiers for both single as well as multi-magnet based DC attraction type levitation system have been discussed and a comparative study has been carried out between these power amplifiers based on their structure, working principle, simulated and experimental results. The effect of different parameters of maglev system on the dynamic responses of amplifiers also been observed.

A cylindrical rod is made to remain suspended under an E-core electromagnet with a single degree of freedom in movement. Modeling, analysis, design, fabrication and testing of the proposed single actuator based DC attraction type levitation system have been done. A comparative study between the performances of different classical controllers for the proposed DCALS at different operating points have been carried out and this study suggests a suitable classical controller that will provide fast response without any overshoots and steady-state error which is generally required for such type of critical systems.

The performances of the linear controller with the change in different parameters of the proposed maglev system have been studied. The parameters of the designed controller also get varied due to many reasons. So a sensitivity study of the controller for a DCALS is also being performed. This study may be useful for designing any corrective measures with parametric uncertainties for such critical systems. The effect of the similar parameters can be studied experimentally and the comparison between the theoretical and practical results would be a future extension of the work.

Design, implementation and testing of a DC attraction type levitation system where a ferromagnetic cylindrical rod of around 0.122 kg mass is made to remain suspended at the desired operating gap under two I-core electromagnets that have been controlled independently through two identical controllers and the stable levitation of the rod is achieved through SISO control of each air-gap corner. The prototype has been successfully tested and stable levitation was demonstrated at the desired operating gap. The use of two actuator based controller demonstrated better pitching movement than single controller based system and by applying suitable vertical gap commands to the two electromagnets, to some extent the roll and heave was also being controlled.

In this work the proposed unstable maglev systems have been stabilized by linear controllers that have designed by classical methods. It has been observed that the closed loop system is stable for the desired operating air-gap but the controllers have a restricted zone of operation. For a small variation of operating air-gap there is sharp degradation of controller performance. But it is essential to design an optimized controller that will stabilize unstable DCALS and will provide satisfactory performance for a wide range of operating air-gap. Two soft-computing tools (i) Genetic Algorithm (GA) and (ii) Particle Swarm Optimisation (PSO) have been utilized for the said purpose. A comparison has been made between the performances of classically designed controllers with GA and PSO based controllers at different operating positions. It has been observed that the performance of PSO based controller is better than GA and classically designed controller.

CONTRIBUTIONS OF THE BOOK

The summary of major contribution of the book is given below.

- Detailed review on DC attraction type levitation system has been carried out.
- Modeling, analysis, design, fabrication and testing of single actuator based DC attraction type levitation system for the suspension of a cylindrical rod have been done.
- Classical controllers have been designed and implemented for the proposed maglev system for different operating points. A comparative study between the performances of controllers has also been investigated both by simulation and experimental results.
- Different possible switched mode power amplifiers for both single as well as multi-magnet based DC attraction type levitation system have been discussed and a comparative study has been carried out between these power amplifiers based on their structure, working principle, simulated and experimental results. The effect of different parameters of maglev system on the dynamic responses of amplifiers also been observed.
- The performance of the proposed control system with the change of different parameters (inductance and resistance of the magnet-coil, input DC link voltage, mass of the payload, air-gap between the pole-face of the magnet and ferromagnetic guide-way etc.) of maglev system has been studied. Sensitivity studies of controller have also been performed.
- Modeling, analysis, design, fabrication and testing of two-magnet based levitation system have been done. Two actuators have been controlled independently through two identical controllers and stable levitation of a cylindrical rod has been achieved through single input and single output (SISO) control of each air-gap utilizing cascade compensation control scheme with inner current loop and outer position loop.
- Optimized controllers have been designed utilizing Genetic Algorithm and Particle Swarm Optimization (PSO) technique for the proposed two actuator based DC attraction type levitation system. It has been

shown that the designed optimized controllers have stabilized unstable DCALS and also performed satisfactorily for a wide range of operating air-gap.

- Simulation results that validate the various controller designs have been presented. The experimental results have also been analyzed to correlate with the theoretical design.

SCOPE OF FUTURE WORK

There are many aspects open for research in the area of DC attraction type levitation systems that may be studied as a future extension of present work. Some of them are listed below.

- The DC attraction type levitation system is an interesting control engineering problem; an unstable plant with changing parameters, nonlinearities and un-modeled dynamics. Other control algorithms may be applied with the existing prototypes to take care of above complexity. Non-linear controllers may be adopted for extending the range of operation. Robust controller may be designed and implemented to compensate modeling error and plant uncertainties of maglev system. Design of an adaptive control scheme for the proposed scheme will also be a good topic of research.
- Implementation of designed optimized controllers for the proposed set-up in dSPACE platform will be an extension of the present work.
- More no of actuators and sensors may be deployed to control movements in other degrees of freedom of the cylindrical rod.
- Air-gap position may be estimated with the change in inductance of magnet-coil and this will lead a position sensor-less levitation system.
- Proposed single switch based power amplifier is not energy efficient. An energy recovery circuit may be used together with the power amplifier for use in multi-actuator based DC attraction type levitation system.

Compilation of References

Abido, M. A. (2002). Optimal design of power-system stabilizers using particle swarm optimization. *IEEE Transactions on Energy Conversion*, *17*(3), 406–413. doi:10.1109/TEC.2002.801992

ANSYS, Inc. (2009). *Licensing guide Release 12.1* (002829). Author.

Banerjee, S., & Bhaduri, R. (2008, December 11–13). The effect of different parameters on the controller performance for a DC attraction type levitation system-a simulation study. In I. I. T. Kanpur (Ed.) *IEEE Conference (INDICON 2008)* (pp. 87–91). 10.1109/INDCON.2008.4768806

Banerjee, S., & Bhaduri, R. (2009). An experimental study towards the selection of operating point for a DC attraction type levitation system. *Modeling. Measurement and Control. Part A*, *64*(1–2), 11–18.

Banerjee, S., Bhaduri, R., & Biswas, P. (2010, April 19–21). *Frequency response based dynamic performance analysis of switched mode power amplifiers used in electromagnetic levitation systems*. In *IEEE International Conference (PEMD 2010)*, Brighton, UK. 10.1049/cp.2010.0183

Banerjee, S., Bhaduri, R., & Prasad, D. (2008). Some aspects of switching power amplifier used in electromagnetic levitation system. *International Journal of Power Electronics*, *1*(1), 33–48. doi:10.1504/IJPELEC.2008.021005

Banerjee, S., Kumar, T. K. S., Pal, J., & Prasad, D. (2008). Controller design for large-gap control of electromagnetically levitated system by using an optimization technique. *IEEE Transactions on Control Systems Technology*, *16*(3), 408–415. doi:10.1109/TCST.2007.906272

Banerjee, S., Prasad, D., & Pal, J. (2006). Large gap control in electromagnetic levitation. *ISA Transactions*, *45*(2), 215–224. doi:10.1016/S0019-0578(07)60191-8 PMID:16649567

Banerjee, S., Sarkar, M. K., Biswas, P. K., Bhaduri, R., & Sarkar, P. (2011). A review note on different components of simple electromagnetic levitation systems. *IETE Technical Review*, *28*(3), 256. Advance online publication. doi:10.4103/0256-4602.81241

Barie, W., & Chiasson, J. (1996). Linear and nonlinear controllers for magnetic levitation. *International Journal of Systems Science*, *27*(11), 1553–1163. doi:10.1080/00207729608929322

Bhaduri, R., Bagchi, S., & Banerjee, S. (2017, April-May). PSO Optimization of the Controller Parameters for an Inherently Magnetic Levitation System. *IACSIT International Journal of Engineering and Technology*, *9*(2), 2319–8613.

Bittar, A., & Sales, R. M. (1997). Control applied to an electromagnetically levitated vehicle. *Proceedings of the IEEE International Conference on Control Applications*, 773–778.

Bittar, A., & Sales, R. M. (1998a). Control for maglev vehicles. *IEEE Control Systems*, *18*(4), 18–25. doi:10.1109/37.710875

Bittar, C. J. J., & Sales, R. M. (1998b). A new approach to the levitation control of an electromagnetic suspension vehicle. *Proceedings of the IEEE International Conference on Control Applications*, 263–267. 10.1109/CCA.1998.728417

Boldea, I., & Nasar, S. A. (1985). *Linear motion electromagnetic systems*. John Wiley & Sons.

Cadence Design Systems. (2000). *Orcad capture user's guide* (2nd ed.).

Carabelli, S., Maddaleno, F., & Muzzarelli, M. (2000). High-efficiency linear power amplifier for active magnetic bearings. *IEEE Transactions on Industrial Electronics*, *47*(1), 17–24. doi:10.1109/41.824021

Chen, J., & Tsukamoto, O. (1991). Attractive type magnetic levitation system with 3 AC electromagnets. *IEEE International Conference on Industrial Electronics, Control y Instrumentation Proceedings*, *1* 778–783.

Cho, D., Kato, Y., & Spilman, D. (1993). Sliding mode and classical controller for magnetic levitation systems. *IEEE Control Systems*, *13*(1), 42–48. doi:10.1109/37.184792

Clark, R. N. (1996). *Control system dynamics*. doi:10.1017/CBO9781139163873

del Valle, Y. D., Venayagamoorthy, G. K., Mohagheghi, S., Hernandez, J.-C., & Harley, R. G. (2008). Particle swarm optimization: Basic concepts, variants and applications in power systems. *IEEE Transactions on Evolutionary Computation*, *12*(2), 171–195. doi:10.1109/TEVC.2007.896686

Eberhart, R. C., & Shi, Y. (1998, May). Comparison between genetic algorithms and particle swarm optimization. In *IEEE International Conference on Evolutionary Computation* (pp. 611–616). IEEE. 10.1007/BFb0040812

Eberhart, R., & Kennedy, J. (1995, October). A new optimizer using particle swarm theory. In *6th International Symposium on Micro Machine and Human Science* (pp. 39–43). 10.1109/MHS.1995.494215

Engelbrecht, A. P. (2006). Particle swarm optimization: Where does it belong? In *Proceedings of the IEEE Swarm Intelligence Symposium* (pp. 48–54). IEEE.

Fogel, D. B. (1995). *Evolutionary computation toward a new philosophy of machine intelligence*. IEEE Publications.

Gaing, Z.-L. (2004). A particle swarm optimization approach for optimum design of PID controller in AVR system. *IEEE Transactions on Energy Conversion, 19*(2), 384–391. doi:10.1109/TEC.2003.821821

Goldberg, D. E. (1989). *Genetic algorithms in search, optimization, and machine learning.* Addison-Wesley Longman Publishing Co., Inc.

Gundogdu, O. (2005). Optimal-tuning of PID controller gains using genetic algorithms. *Journal of Engineering Sciences, 11*(1), 131–135.

Hu, X. (2006). *Particle swarm optimization.* Tutorial of the IEEE Swarm Intelligence Symposium.

Hurley, W., & Wolfle, W. (1997). Electromagnetic design of a magnetic suspension system. *IEEE Transactions on Education, 40*(2), 124–130. doi:10.1109/13.572325

Jan, J. A., & Sulc, B. (2002, May 27–30). Evolutionary computing methods for optimizing virtual reality process models. In *International Carpathian Control. Conference (ICCC 2002).* Malenovice.

Jayawant, B. V. (1988). Review lecture on electromagnetic suspension and levitation techniques. *Proceedings of the Royal Society of London. Series A, Mathematical and Physical Sciences, 416*(1851), 245–320. doi:10.1098/rspa.1988.0036

Jayawant, B. V., Sinha, P. K., Wheeler, A. R., Whorlow, R. J., & Willsher, J. (1976). Development of 1-ton magnetically suspended vehicle using controlled d.c. electromagnets. *Proceedings of the Institution of Electrical Engineers, 123*(9), 941–948. doi:10.1049/piee.1976.0203

Joo, S., & Seo, J. H. (1997). Design and analysis of the nonlinear feedback linearizing control for an electromagnetic suspension system. *IEEE Transactions on Control Systems Technology, 5*(1), 593–598.

Kennedy, J., & Eberhart, R. (1995, November). Particle swarm optimization. In *IEEE International Conference on Neural Networks* (pp. 1942–1948). IEEE.

Kim, D. H., & Park, J. I., III. (2005). Intelligent PID controller tuning of AVR system using GA and PSO. Lecture Notes in Computer Science, 375. doi:10.1007/11538356_38

Kim, Y. C. (1994). Gain scheduled control of magnetic suspension system. In *Proceedings of the American Control Conference* (pp. 3127–3131). Academic Press.

Krishnan, R. (2003). *Electric motor drives: Modeling, analysis, and control.* Prentice Hall of India.

Krohling, R. A., Jaschek, H., & Rey, J. P. (1997). Designing PI/PID controller for a motion control system based on genetic algorithm. In *Proceedings of the 12th IEEE International Symposium on Intelligent Control* (pp. 125–130). 10.1109/ISIC.1997.626429

Lin, F. J., Chen, S. Y., & Shyu, K. K. (2009). Robust dynamic sliding-mode control using adaptive RENN for magnetic levitation system. *IEEE Transactions on Neural Networks, 20*(6), 938–951. doi:10.1109/TNN.2009.2014228 PMID:19423437

Lin, J., & Tho, B. (1998). Analysis and -Based controller design for an electromagnetic suspension system. *IEEE Transactions on Education, 41*(2), 116–129. doi:10.1109/13.669722

Maddaleno, F., Carabelli, S., & Muzzarelli, M. (1997). A modified class G amplifier for active magnetic bearings. *13th Annual Applied Power Electronics Conference Proceedings, 1*, 534–538.

Maddaleno, F., Carabelli, S., & Muzzarelli, M. (1997). A modified class G amplifier for active magnetic bearings. *13th Annual Applied Power Electronics Conference Proceedings, 1*, 534–538. 10.1109/APEC.1997.581510

Mahony, T. O., Downing, C. J., & Fatla, K. (2000). Genetic algorithm for PID parameter optimization: Minimizing error criteria. *Process Control. and Instrumentation*, 148–153.

Man, K. F., Tang, K. S., & Kwong, S. (1996). Genetic algorithms: Concepts and applications. *IEEE Transactions on Industrial Electronics, 43*(5), 519–534. doi:10.1109/41.538609

Mohan, N., Undeland, T. M., & Robbins, W. P. (1996). *Power electronic converters: Applications and design* (2nd ed.). IEEE Publications and John Wiley & Sons.

Moon, F. C. (1994). *Superconducting levitation, applications to bearings and magnetic transportation*. John Wiley & Sons.

Moriyama, S. (2000). AC magnetic suspension using a differential feedback power amplifier. *Electrical Engineering in Japan, 131*(2), 61–67. doi:10.1002/(SICI)1520-6416(20000430)131:2<61::AID-EEJ7>3.0.CO;2-Q

Namerikawa, T., & Fujita, M. (2001). Uncertainty structure and μ-synthesis of a magnetic suspension system. *IEEJ Transactions on Electronics. Information Systems, 121-C*(6), 1080–1087. doi:10.1541/ieejeiss1987.121.6_1080

Nasri, M., Nezamabadi-pour, H., & Maghfoori, M. (2007). A PSO-based optimum design of PID controller for a linear brushless DC motor. *World Academy of Science, Engineering and Technology*, 211–215.

Ogata, K. (2000). *Modern control engineering* (3rd ed.). Prentice Hall of India.

Oliveira, V. A., Costa, E. F., & Vargas, J. B. (1999). Digital implementation of a magnetic suspension system for laboratory experiments. *IEEE Transactions on Education, 42*(4), 315–322. doi:10.1109/13.804538

Rathikarani, D., Sivakumar, D., & Mary, S. A. J. (2007). Genetic Algorithm based Optimization of Controller Tuning for Processes with Dead Time. Advances in Control and Optimization of Dynamical Systems.

Rogg, D. (1984). General survey of the possible applications and development tendencies of magnetic levitation technology. *IEEE Transactions on Magnetics, 20*(5), 1696–1701. doi:10.1109/TMAG.1984.1063347

Sadasivarao, M. V., & Chidambaram, M. (2006). PID controller tuning of cascade control systems using genetic algorithm. *Journal of the Indian Institute of Science*, 343–354.

Salami, M., & Cain, G. (1995). An adaptive PID controller based on genetic algorithm processor. Genetic Algorithms in Engineering Systems, (414). doi:10.1049/cp:19951030

Shiao, Y. S. (2001). Design and implementation of a controller for a magnetic levitation system. *Proceedings of the National Science Council, 11*(2), 88–94.

Shibukawa, K., Tsubakiji, T., & Kimura, H. (1991). Robust stabilization of a magnetic levitation system. *Proceedings of the 30th Conference on Decision and Control*, 2368–2371. 10.1109/CDC.1991.261609

Shin, G. W., Song, Y. J., Lee, T. B., & Choi, H. K. (2007). Genetic algorithm for identification of time delay systems from step responses. *International Journal of Control, Automation, and Systems*, *5*(1), 79–85.

Shirazee, N. A., & Basak, A. (1995). Electro permanent suspension system for acquiring large air gaps to suspend loads. *IEEE Transactions on Magnetics*, *31*(6), 4193–4195. doi:10.1109/20.489923

Shi, Y., & Eberhart, R. (1998a). A modified particle swarm optimizer. In *Proceedings of the IEEE International Conference on Evolutionary Computation* (pp. 69–73). IEEE Publications.

Shi, Y., & Eberhart, R. (1998b). Parameter selection in particle swarm optimization. In *Proceedings of the 7th International Conference on Evolutionary Programming (EP)* (pp. 591–600). IEEE.

Shi, Y., & Eberhart, R. C. (1999c). Empirical study of particle swarm optimization. In *Proceedings of the IEEE Congress on Evolutionary Computation* (pp. 1945–1950). IEEE. 10.1109/CEC.1999.785511

Sinha, P. K. (1987). *Electromagnetic suspension, dynamics and control. Peter Peregrinus Ltd.*

Trumper, D. L., Olson, S. M., & Subrahmanyan, P. K. (1997). Linearizing control of magnetic suspension systems. *IEEE Transactions on Control Systems Technology*, *5*(4), 427–438. doi:10.1109/87.595924

Tsukamoto, O., Yasuda, K., & Chen, J. Z. (1988). A new magnetic levitation system with AC magnets. *IEEE Transactions on Magnetics*, *24*(2), 1497–1500. doi:10.1109/20.11528

Wong, T. H. (1986). Design of a magnetic levitation control system – An undergraduate project. *IEEE Transactions on Education*, *29*(4), 196–200. doi:10.1109/TE.1986.5570565

Yamamura, S. (1976). Magnetic levitation technology of tracked vehicles: Present status and prospects. *IEEE Transactions on Magnetics*, *12*(6), 874–878. doi:10.1109/TMAG.1976.1059125

Yoshida, H., Kawata, K., Fukuyama, Y., Takayama, S., & Nakanishi, Y. (2000). A particle swarm optimization for reactive power and voltage control considering voltage security assessment. *IEEE Transactions on Power Systems*, *15*(4), 1232–1239. doi:10.1109/59.898095

Related References

To continue our tradition of advancing academic research, we have compiled a list of recommended IGI Global readings. These references will provide additional information and guidance to further enrich your knowledge and assist you with your own research and future publications.

Abbasnejad, B., Moeinzadeh, S., Ahankoob, A., & Wong, P. S. (2021). The Role of Collaboration in the Implementation of BIM-Enabled Projects. In J. Underwood & M. Shelbourn (Eds.), *Handbook of Research on Driving Transformational Change in the Digital Built Environment* (pp. 27–62). IGI Global. https://doi.org/10.4018/978-1-7998-6600-8.ch002

Abdulrahman, K. O., Mahamood, R. M., & Akinlabi, E. T. (2022). Additive Manufacturing (AM): Processing Technique for Lightweight Alloys and Composite Material. In K. Kumar, B. Babu, & J. Davim (Ed.), *Handbook of Research on Advancements in the Processing, Characterization, and Application of Lightweight Materials* (pp. 27-48). IGI Global. https://doi.org/10.4018/978-1-7998-7864-3.ch002

Agrawal, R., Sharma, P., & Saxena, A. (2021). A Diamond Cut Leather Substrate Antenna for BAN (Body Area Network) Application. In V. Singh, V. Dubey, A. Saxena, R. Tiwari, & H. Sharma (Eds.), *Emerging Materials and Advanced Designs for Wearable Antennas* (pp. 54–59). IGI Global. https://doi.org/10.4018/978-1-7998-7611-3.ch004

Ahmad, F., Al-Ammar, E. A., & Alsaidan, I. (2022). Battery Swapping Station: A Potential Solution to Address the Limitations of EV Charging Infrastructure. In M. Alam, R. Pillai, & N. Murugesan (Eds.), *Developing Charging Infrastructure and Technologies for Electric Vehicles* (pp. 195–207). IGI Global. doi:10.4018/978-1-7998-6858-3.ch010

Aikhuele, D. (2018). A Study of Product Development Engineering and Design Reliability Concerns. *International Journal of Applied Industrial Engineering*, *5*(1), 79–89. doi:10.4018/IJAIE.2018010105

Al-Khatri, H., & Al-Atrash, F. (2021). Occupants' Habits and Natural Ventilation in a Hot Arid Climate. In R. González-Lezcano (Ed.), *Advancements in Sustainable Architecture and Energy Efficiency* (pp. 146–168). IGI Global. https://doi.org/10.4018/978-1-7998-7023-4.ch007

Al-Shebeeb, O. A., Rangaswamy, S., Gopalakrishan, B., & Devaru, D. G. (2017). Evaluation and Indexing of Process Plans Based on Electrical Demand and Energy Consumption. *International Journal of Manufacturing, Materials, and Mechanical Engineering*, *7*(3), 1–19. doi:10.4018/IJMMME.2017070101

Amuda, M. O., Lawal, T. F., & Akinlabi, E. T. (2017). Research Progress on Rheological Behavior of AA7075 Aluminum Alloy During Hot Deformation. *International Journal of Materials Forming and Machining Processes*, *4*(1), 53–96. doi:10.4018/IJMFMP.2017010104

Amuda, M. O., Lawal, T. F., & Mridha, S. (2021). Microstructure and Mechanical Properties of Silicon Carbide-Treated Ferritic Stainless Steel Welds. In L. Burstein (Ed.), *Handbook of Research on Advancements in Manufacturing, Materials, and Mechanical Engineering* (pp. 395–411). IGI Global. https://doi.org/10.4018/978-1-7998-4939-1.ch019

Anikeev, V., Gasem, K. A., & Fan, M. (2021). Application of Supercritical Technologies in Clean Energy Production: A Review. In L. Chen (Ed.), *Handbook of Research on Advancements in Supercritical Fluids Applications for Sustainable Energy Systems* (pp. 792–821). IGI Global. https://doi.org/10.4018/978-1-7998-5796-9.ch022

Arafat, M. Y., Saleem, I., & Devi, T. P. (2022). Drivers of EV Charging Infrastructure Entrepreneurship in India. In M. Alam, R. Pillai, & N. Murugesan (Eds.), *Developing Charging Infrastructure and Technologies for Electric Vehicles* (pp. 208–219). IGI Global. https://doi.org/10.4018/978-1-7998-6858-3.ch011

Araujo, A., & Manninen, H. (2022). Contribution of Project-Based Learning on Social Skills Development: An Industrial Engineer Perspective. In A. Alves & N. van Hattum-Janssen (Eds.), *Training Engineering Students for Modern Technological Advancement* (pp. 119–145). IGI Global. https://doi.org/10.4018/978-1-7998-8816-1.ch006

Armutlu, H. (2018). Intelligent Biomedical Engineering Operations by Cloud Computing Technologies. In U. Kose, G. Guraksin, & O. Deperlioglu (Eds.), *Nature-Inspired Intelligent Techniques for Solving Biomedical Engineering Problems* (pp. 297–317). Hershey, PA: IGI Global. doi:10.4018/978-1-5225-4769-3.ch015

Atik, M., Sadek, M., & Shahrour, I. (2017). Single-Run Adaptive Pushover Procedure for Shear Wall Structures. In V. Plevris, G. Kremmyda, & Y. Fahjan (Eds.), *Performance-Based Seismic Design of Concrete Structures and Infrastructures* (pp. 59–83). Hershey, PA: IGI Global. doi:10.4018/978-1-5225-2089-4.ch003

Attia, H. (2021). Smart Power Microgrid Impact on Sustainable Building. In R. González-Lezcano (Ed.), *Advancements in Sustainable Architecture and Energy Efficiency* (pp. 169–194). IGI Global. https://doi.org/10.4018/978-1-7998-7023-4.ch008

Aydin, A., Akyol, E., Gungor, M., Kaya, A., & Tasdelen, S. (2018). Geophysical Surveys in Engineering Geology Investigations With Field Examples. In N. Ceryan (Ed.), *Handbook of Research on Trends and Digital Advances in Engineering Geology* (pp. 257–280). Hershey, PA: IGI Global. doi:10.4018/978-1-5225-2709-1.ch007

Ayoobkhan, M. U. D., Y., A., J., Easwaran, B., & R., T. (2021). Smart Connected Digital Products and IoT Platform With the Digital Twin. In P. Vasant, G. Weber, & W. Punurai (Ed.), Research Advancements in Smart Technology, Optimization, and Renewable Energy (pp. 330-350). IGI Global. https://doi.org/ doi:10.4018/978-1-7998-3970-5.ch016

Baeza Moyano, D., & González Lezcano, R. A. (2021). The Importance of Light in Our Lives: Towards New Lighting in Schools. In R. González-Lezcano (Ed.), *Advancements in Sustainable Architecture and Energy Efficiency* (pp. 239–256). IGI Global. https://doi.org/10.4018/978-1-7998-7023-4.ch011

Bagdadee, A. H. (2021). A Brief Assessment of the Energy Sector of Bangladesh. *International Journal of Energy Optimization and Engineering*, *10*(1), 36–55. doi:10.4018/IJEOE.2021010103

Baklezos, A. T., & Hadjigeorgiou, N. G. (2021). Magnetic Sensors for Space Applications and Magnetic Cleanliness Considerations. In C. Nikolopoulos (Ed.), *Recent Trends on Electromagnetic Environmental Effects for Aeronautics and Space Applications* (pp. 147–185). IGI Global. https://doi.org/10.4018/978-1-7998-4879-0.ch006

Bas, T. G. (2017). Nutraceutical Industry with the Collaboration of Biotechnology and Nutrigenomics Engineering: The Significance of Intellectual Property in the Entrepreneurship and Scientific Research Ecosystems. In T. Bas & J. Zhao (Eds.), *Comparative Approaches to Biotechnology Development and Use in Developed and Emerging Nations* (pp. 1–17). Hershey, PA: IGI Global. doi:10.4018/978-1-5225-1040-6.ch001

Bazeer Ahamed, B., & Periakaruppan, S. (2021). Taxonomy of Influence Maximization Techniques in Unknown Social Networks. In P. Vasant, G. Weber, & W. Punurai (Eds.), *Research Advancements in Smart Technology, Optimization, and Renewable Energy* (pp. 351-363). IGI Global. https://doi.org/10.4018/978-1-7998-3970-5.ch017

Beale, R., & André, J. (2017). *Design Solutions and Innovations in Temporary Structures*. Hershey, PA: IGI Global. doi:10.4018/978-1-5225-2199-0

Behnam, B. (2017). Simulating Post-Earthquake Fire Loading in Conventional RC Structures. In P. Samui, S. Chakraborty, & D. Kim (Eds.), *Modeling and Simulation Techniques in Structural Engineering* (pp. 425–444). Hershey, PA: IGI Global. doi:10.4018/978-1-5225-0588-4.ch015

Ben Hamida, I., Salah, S. B., Msahli, F., & Mimouni, M. F. (2018). Distribution Network Reconfiguration Using SPEA2 for Power Loss Minimization and Reliability Improvement. *International Journal of Energy Optimization and Engineering*, *7*(1), 50–65. doi:10.4018/IJEOE.2018010103

Bentarzi, H. (2021). Fault Tree-Based Root Cause Analysis Used to Study Mal-Operation of a Protective Relay in a Smart Grid. In A. Recioui & H. Bentarzi (Eds.), *Optimizing and Measuring Smart Grid Operation and Control* (pp. 289–308). IGI Global. https://doi.org/10.4018/978-1-7998-4027-5.ch012

Beysens, D. A., Garrabos, Y., & Zappoli, B. (2021). Thermal Effects in Near-Critical Fluids: Piston Effect and Related Phenomena. In L. Chen (Ed.), *Handbook of Research on Advancements in Supercritical Fluids Applications for Sustainable Energy Systems* (pp. 1–31). IGI Global. https://doi.org/10.4018/978-1-7998-5796-9.ch001

Bhaskar, S. V., & Kudal, H. N. (2017). Effect of TiCN and AlCrN Coating on Tribological Behaviour of Plasma-nitrided AISI 4140 Steel. *International Journal of Surface Engineering and Interdisciplinary Materials Science*, *5*(2), 1–17. doi:10.4018/IJSEIMS.2017070101

Bhuyan, D. (2018). Designing of a Twin Tube Shock Absorber: A Study in Reverse Engineering. In K. Kumar & J. Davim (Eds.), *Design and Optimization of Mechanical Engineering Products* (pp. 83–104). Hershey, PA: IGI Global. doi:10.4018/978-1-5225-3401-3.ch005

Blumberg, G. (2021). Blockchains for Use in Construction and Engineering Projects. In J. Underwood & M. Shelbourn (Eds.), *Handbook of Research on Driving Transformational Change in the Digital Built Environment* (pp. 179–208). IGI Global. https://doi.org/10.4018/978-1-7998-6600-8.ch008

Bolboaca, A. M. (2021). Considerations Regarding the Use of Fuel Cells in Combined Heat and Power for Stationary Applications. In G. Badea, R. Felseghi, & I. Aşchilean (Eds.), *Hydrogen Fuel Cell Technology for Stationary Applications* (pp. 239–275). IGI Global. https://doi.org/10.4018/978-1-7998-4945-2.ch010

Burstein, L. (2021). Simulation Tool for Cable Design. In L. Burstein (Ed.), *Handbook of Research on Advancements in Manufacturing, Materials, and Mechanical Engineering* (pp. 54–74). IGI Global. https://doi.org/10.4018/978-1-7998-4939-1.ch003

Calderon, F. A., Giolo, E. G., Frau, C. D., Rengel, M. G., Rodriguez, H., Tornello, M., ... Gallucci, R. (2018). Seismic Microzonation and Site Effects Detection Through Microtremors Measures: A Review. In N. Ceryan (Ed.), *Handbook of Research on Trends and Digital Advances in Engineering Geology* (pp. 326–349). Hershey, PA: IGI Global. doi:10.4018/978-1-5225-2709-1.ch009

Ceryan, N., & Can, N. K. (2018). Prediction of The Uniaxial Compressive Strength of Rocks Materials. In N. Ceryan (Ed.), *Handbook of Research on Trends and Digital Advances in Engineering Geology* (pp. 31–96). Hershey, PA: IGI Global. doi:10.4018/978-1-5225-2709-1.ch002

Ceryan, S. (2018). Weathering Indices Used in Evaluation of the Weathering State of Rock Material. In N. Ceryan (Ed.), *Handbook of Research on Trends and Digital Advances in Engineering Geology* (pp. 132–186). Hershey, PA: IGI Global. doi:10.4018/978-1-5225-2709-1.ch004

Chen, H., Padilla, R. V., & Besarati, S. (2017). Supercritical Fluids and Their Applications in Power Generation. In L. Chen & Y. Iwamoto (Eds.), *Advanced Applications of Supercritical Fluids in Energy Systems* (pp. 369–402). Hershey, PA: IGI Global. doi:10.4018/978-1-5225-2047-4.ch012

Chen, H., Padilla, R. V., & Besarati, S. (2021). Supercritical Fluids and Their Applications in Power Generation. In L. Chen (Ed.), *Handbook of Research on Advancements in Supercritical Fluids Applications for Sustainable Energy Systems* (pp. 566–599). IGI Global. https://doi.org/10.4018/978-1-7998-5796-9.ch016

Chen, L. (2017). Principles, Experiments, and Numerical Studies of Supercritical Fluid Natural Circulation System. In L. Chen & Y. Iwamoto (Eds.), *Advanced Applications of Supercritical Fluids in Energy Systems* (pp. 136–187). Hershey, PA: IGI Global. doi:10.4018/978-1-5225-2047-4.ch005

Chen, L. (2021). Principles, Experiments, and Numerical Studies of Supercritical Fluid Natural Circulation System. In L. Chen (Ed.), *Handbook of Research on Advancements in Supercritical Fluids Applications for Sustainable Energy Systems* (pp. 219–269). IGI Global. https://doi.org/10.4018/978-1-7998-5796-9.ch007

Chiba, Y., Marif, Y., Henini, N., & Tlemcani, A. (2021). Modeling of Magnetic Refrigeration Device by Using Artificial Neural Networks Approach. *International Journal of Energy Optimization and Engineering*, *10*(4), 68–76. https://doi.org/10.4018/IJEOE.2021100105

Clementi, F., Di Sciascio, G., Di Sciascio, S., & Lenci, S. (2017). Influence of the Shear-Bending Interaction on the Global Capacity of Reinforced Concrete Frames: A Brief Overview of the New Perspectives. In V. Plevris, G. Kremmyda, & Y. Fahjan (Eds.), *Performance-Based Seismic Design of Concrete Structures and Infrastructures* (pp. 84–111). Hershey, PA: IGI Global. doi:10.4018/978-1-5225-2089-4.ch004

Codinhoto, R., Fialho, B. C., Pinti, L., & Fabricio, M. M. (2021). BIM and IoT for Facilities Management: Understanding Key Maintenance Issues. In J. Underwood & M. Shelbourn (Eds.), *Handbook of Research on Driving Transformational Change in the Digital Built Environment* (pp. 209–231). IGI Global. doi:10.4018/978-1-7998-6600-8.ch009

Cortés-Polo, D., Calle-Cancho, J., Carmona-Murillo, J., & González-Sánchez, J. (2017). Future Trends in Mobile-Fixed Integration for Next Generation Networks: Classification and Analysis. *International Journal of Vehicular Telematics and Infotainment Systems, 1*(1), 33–53. doi:10.4018/IJVTIS.2017010103

Costa, H. G., Sheremetieff, F. H., & Araújo, E. A. (2022). Influence of Game-Based Methods in Developing Engineering Competences. In A. Alves & N. van Hattum-Janssen (Eds.), *Training Engineering Students for Modern Technological Advancement* (pp. 69–88). IGI Global. https://doi.org/10.4018/978-1-7998-8816-1.ch004

Cui, X., Zeng, S., Li, Z., Zheng, Q., Yu, X., & Han, B. (2018). Advanced Composites for Civil Engineering Infrastructures. In K. Kumar & J. Davim (Eds.), *Composites and Advanced Materials for Industrial Applications* (pp. 212–248). Hershey, PA: IGI Global. doi:10.4018/978-1-5225-5216-1.ch010

Dalgıç, S., & Kuşku, İ. (2018). Geological and Geotechnical Investigations in Tunneling. In N. Ceryan (Ed.), *Handbook of Research on Trends and Digital Advances in Engineering Geology* (pp. 482–529). Hershey, PA: IGI Global. doi:10.4018/978-1-5225-2709-1.ch014

Dang, C., & Hihara, E. (2021). Study on Cooling Heat Transfer of Supercritical Carbon Dioxide Applied to Transcritical Carbon Dioxide Heat Pump. In L. Chen (Ed.), *Handbook of Research on Advancements in Supercritical Fluids Applications for Sustainable Energy Systems* (pp. 451–493). IGI Global. https://doi.org/10.4018/978-1-7998-5796-9.ch013

Daus, Y., Kharchenko, V., & Yudaev, I. (2021). Research of Solar Energy Potential of Photovoltaic Installations on Enclosing Structures of Buildings. *International Journal of Energy Optimization and Engineering*, *10*(4), 18–34. https://doi.org/10.4018/IJEOE.2021100102

Daus, Y., Kharchenko, V., & Yudaev, I. (2021). Optimizing Layout of Distributed Generation Sources of Power Supply System of Agricultural Object. *International Journal of Energy Optimization and Engineering*, *10*(3), 70–84. https://doi.org/10.4018/IJEOE.2021070104

de la Varga, D., Soto, M., Arias, C. A., van Oirschot, D., Kilian, R., Pascual, A., & Álvarez, J. A. (2017). Constructed Wetlands for Industrial Wastewater Treatment and Removal of Nutrients. In Á. Val del Río, J. Campos Gómez, & A. Mosquera Corral (Eds.), *Technologies for the Treatment and Recovery of Nutrients from Industrial Wastewater* (pp. 202–230). Hershey, PA: IGI Global. doi:10.4018/978-1-5225-1037-6.ch008

Deb, S., Ammar, E. A., AlRajhi, H., Alsaidan, I., & Shariff, S. M. (2022). V2G Pilot Projects: Review and Lessons Learnt. In M. Alam, R. Pillai, & N. Murugesan (Eds.), *Developing Charging Infrastructure and Technologies for Electric Vehicles* (pp. 252–267). IGI Global. https://doi.org/10.4018/978-1-7998-6858-3.ch014

Dekhandji, F. Z., & Rais, M. C. (2021). A Comparative Study of Power Quality Monitoring Using Various Techniques. In A. Recioui & H. Bentarzi (Eds.), *Optimizing and Measuring Smart Grid Operation and Control* (pp. 259–288). IGI Global. https://doi.org/10.4018/978-1-7998-4027-5.ch011

Deperlioglu, O. (2018). Intelligent Techniques Inspired by Nature and Used in Biomedical Engineering. In U. Kose, G. Guraksin, & O. Deperlioglu (Eds.), *Nature-Inspired Intelligent Techniques for Solving Biomedical Engineering Problems* (pp. 51–77). Hershey, PA: IGI Global. doi:10.4018/978-1-5225-4769-3.ch003

Dhurpate, P. R., & Tang, H. (2021). Quantitative Analysis of the Impact of Inter-Line Conveyor Capacity for Throughput of Manufacturing Systems. *International Journal of Manufacturing, Materials, and Mechanical Engineering*, *11*(1), 1–17. https://doi.org/10.4018/IJMMME.2021010101

Dinkar, S., & Deep, K. (2021). A Survey of Recent Variants and Applications of Antlion Optimizer. *International Journal of Energy Optimization and Engineering*, *10*(2), 48–73. doi:10.4018/IJEOE.2021040103

Dixit, A. (2018). Application of Silica-Gel-Reinforced Aluminium Composite on the Piston of Internal Combustion Engine: Comparative Study of Silica-Gel-Reinforced Aluminium Composite Piston With Aluminium Alloy Piston. In K. Kumar & J. Davim (Eds.), *Composites and Advanced Materials for Industrial Applications* (pp. 63–98). Hershey, PA: IGI Global. doi:10.4018/978-1-5225-5216-1.ch004

Drabecki, M. P., & Kułak, K. B. (2021). Global Pandemics on European Electrical Energy Markets: Lessons Learned From the COVID-19 Outbreak. *International Journal of Energy Optimization and Engineering*, *10*(3), 24–46. https://doi.org/10.4018/IJEOE.2021070102

Dutta, M. M. (2021). Nanomaterials for Food and Agriculture. In M. Bhat, I. Wani, & S. Ashraf (Eds.), *Applications of Nanomaterials in Agriculture, Food Science, and Medicine* (pp. 75–97). IGI Global. doi:10.4018/978-1-7998-5563-7.ch004

Dutta, M. M., & Goswami, M. (2021). Coating Materials: Nano-Materials. In S. Roy & G. Bose (Eds.), *Advanced Surface Coating Techniques for Modern Industrial Applications* (pp. 1–30). IGI Global. doi:10.4018/978-1-7998-4870-7.ch001

Elsayed, A. M., Dakkama, H. J., Mahmoud, S., Al-Dadah, R., & Kaialy, W. (2017). Sustainable Cooling Research Using Activated Carbon Adsorbents and Their Environmental Impact. In T. Kobayashi (Ed.), *Applied Environmental Materials Science for Sustainability* (pp. 186–221). Hershey, PA: IGI Global. doi:10.4018/978-1-5225-1971-3.ch009

Ercanoglu, M., & Sonmez, H. (2018). General Trends and New Perspectives on Landslide Mapping and Assessment Methods. In N. Ceryan (Ed.), *Handbook of Research on Trends and Digital Advances in Engineering Geology* (pp. 350–379). Hershey, PA: IGI Global. doi:10.4018/978-1-5225-2709-1.ch010

Faroz, S. A., Pujari, N. N., Rastogi, R., & Ghosh, S. (2017). Risk Analysis of Structural Engineering Systems Using Bayesian Inference. In P. Samui, S. Chakraborty, & D. Kim (Eds.), *Modeling and Simulation Techniques in Structural Engineering* (pp. 390–424). Hershey, PA: IGI Global. doi:10.4018/978-1-5225-0588-4.ch014

Fekik, A., Hamida, M. L., Denoun, H., Azar, A. T., Kamal, N. A., Vaidyanathan, S., Bousbaine, A., & Benamrouche, N. (2022). Multilevel Inverter for Hybrid Fuel Cell/PV Energy Conversion System. In A. Fekik & N. Benamrouche (Eds.), *Modeling and Control of Static Converters for Hybrid Storage Systems* (pp. 233–270). IGI Global. https://doi.org/10.4018/978-1-7998-7447-8.ch009

Fekik, A., Hamida, M. L., Houassine, H., Azar, A. T., Kamal, N. A., Denoun, H., Vaidyanathan, S., & Sambas, A. (2022). Power Quality Improvement for Grid-Connected Photovoltaic Panels Using Direct Power Control. In A. Fekik & N. Benamrouche (Eds.), *Modeling and Control of Static Converters for Hybrid Storage Systems* (pp. 107–142). IGI Global. https://doi.org/10.4018/978-1-7998-7447-8.ch005

Fernando, P. R., Hamigah, T., Disne, S., Wickramasingha, G. G., & Sutharshan, A. (2018). The Evaluation of Engineering Properties of Low Cost Concrete Blocks by Partial Doping of Sand with Sawdust: Low Cost Sawdust Concrete Block. *International Journal of Strategic Engineering*, *1*(2), 26–42. doi:10.4018/IJoSE.2018070103

Ferro, G., Minciardi, R., Parodi, L., & Robba, M. (2022). Optimal Charging Management of Microgrid-Integrated Electric Vehicles. In M. Alam, R. Pillai, & N. Murugesan (Eds.), *Developing Charging Infrastructure and Technologies for Electric Vehicles* (pp. 133–155). IGI Global. https://doi.org/10.4018/978-1-7998-6858-3.ch007

Flumerfelt, S., & Green, C. (2022). Graduate Lean Leadership Education: A Case Study of a Program. In A. Alves & N. van Hattum-Janssen (Eds.), *Training Engineering Students for Modern Technological Advancement* (pp. 202–224). IGI Global. https://doi.org/10.4018/978-1-7998-8816-1.ch010

Galli, B. J. (2021). Implications of Economic Decision Making to the Project Manager. *International Journal of Strategic Engineering*, *4*(1), 19–32. https://doi.org/10.4018/IJoSE.2021010102

Gento, A. M., Pimentel, C., & Pascual, J. A. (2022). Teaching Circular Economy and Lean Management in a Learning Factory. In A. Alves & N. van Hattum-Janssen (Eds.), *Training Engineering Students for Modern Technological Advancement* (pp. 183–201). IGI Global. https://doi.org/10.4018/978-1-7998-8816-1.ch009

Ghosh, S., Mitra, S., Ghosh, S., & Chakraborty, S. (2017). Seismic Reliability Analysis in the Framework of Metamodelling Based Monte Carlo Simulation. In P. Samui, S. Chakraborty, & D. Kim (Eds.), *Modeling and Simulation Techniques in Structural Engineering* (pp. 192–208). Hershey, PA: IGI Global. doi:10.4018/978-1-5225-0588-4.ch006

Gil, M., & Otero, B. (2017). Learning Engineering Skills through Creativity and Collaboration: A Game-Based Proposal. In R. Alexandre Peixoto de Queirós & M. Pinto (Eds.), *Gamification-Based E-Learning Strategies for Computer Programming Education* (pp. 14–29). Hershey, PA: IGI Global. doi:10.4018/978-1-5225-1034-5.ch002

Gill, J., Ayre, M., & Mills, J. (2017). Revisioning the Engineering Profession: How to Make It Happen! In M. Gray & K. Thomas (Eds.), *Strategies for Increasing Diversity in Engineering Majors and Careers* (pp. 156–175). Hershey, PA: IGI Global. doi:10.4018/978-1-5225-2212-6.ch008

Godzhaev, Z., Senkevich, S., Kuzmin, V., & Melikov, I. (2021). Use of the Neural Network Controller of Sprung Mass to Reduce Vibrations From Road Irregularities. In P. Vasant, G. Weber, & W. Punurai (Ed.), *Research Advancements in Smart Technology, Optimization, and Renewable Energy* (pp. 69-87). IGI Global. https://doi.org/10.4018/978-1-7998-3970-5.ch005

Gomes de Gusmão, C. M. (2022). Digital Competencies and Transformation in Higher Education: Upskilling With Extension Actions. In A. Alves & N. van Hattum-Janssen (Eds.), *Training Engineering Students for Modern Technological Advancement* (pp. 313–328). IGI Global. https://doi.org/10.4018/978-1-7998-8816-1.ch015A

Goyal, N., Ram, M., & Kumar, P. (2017). Welding Process under Fault Coverage Approach for Reliability and MTTF. In M. Ram & J. Davim (Eds.), *Mathematical Concepts and Applications in Mechanical Engineering and Mechatronics* (pp. 222–245). Hershey, PA: IGI Global. doi:10.4018/978-1-5225-1639-2.ch011

Gray, M., & Lundy, C. (2017). Engineering Study Abroad: High Impact Strategy for Increasing Access. In M. Gray & K. Thomas (Eds.), *Strategies for Increasing Diversity in Engineering Majors and Careers* (pp. 42–59). Hershey, PA: IGI Global. doi:10.4018/978-1-5225-2212-6.ch003

Güler, O., & Varol, T. (2021). Fabrication of Functionally Graded Metal and Ceramic Powders Synthesized by Electroless Deposition. In S. Roy & G. Bose (Eds.), *Advanced Surface Coating Techniques for Modern Industrial Applications* (pp. 150–187). IGI Global. https://doi.org/10.4018/978-1-7998-4870-7.ch007

Guraksin, G. E. (2018). Internet of Things and Nature-Inspired Intelligent Techniques for the Future of Biomedical Engineering. In U. Kose, G. Guraksin, & O. Deperlioglu (Eds.), *Nature-Inspired Intelligent Techniques for Solving Biomedical Engineering Problems* (pp. 263–282). Hershey, PA: IGI Global. doi:10.4018/978-1-5225-4769-3.ch013

Hamida, M. L., Fekik, A., Denoun, H., Ardjal, A., & Bokhtache, A. A. (2022). Flying Capacitor Inverter Integration in a Renewable Energy System. In A. Fekik & N. Benamrouche (Eds.), *Modeling and Control of Static Converters for Hybrid Storage Systems* (pp. 287–306). IGI Global. https://doi.org/10.4018/978-1-7998-7447-8.ch011

Hasegawa, N., & Takahashi, Y. (2021). Control of Soap Bubble Ejection Robot Using Facial Expressions. *International Journal of Manufacturing, Materials, and Mechanical Engineering, 11*(2), 1–16. https://doi.org/10.4018/IJMMME.2021040101

Hejazi, T., & Akbari, L. (2017). A Multiresponse Optimization Model for Statistical Design of Processes with Discrete Variables. In M. Ram & J. Davim (Eds.), *Mathematical Concepts and Applications in Mechanical Engineering and Mechatronics* (pp. 17–37). Hershey, PA: IGI Global. doi:10.4018/978-1-5225-1639-2.ch002

Hejazi, T., & Hejazi, A. (2017). Monte Carlo Simulation for Reliability-Based Design of Automotive Complex Subsystems. In M. Ram & J. Davim (Eds.), *Mathematical Concepts and Applications in Mechanical Engineering and Mechatronics* (pp. 177–200). Hershey, PA: IGI Global. doi:10.4018/978-1-5225-1639-2.ch009

Hejazi, T., & Poursabbagh, H. (2017). Reliability Analysis of Engineering Systems: An Accelerated Life Testing for Boiler Tubes. In M. Ram & J. Davim (Eds.), *Mathematical Concepts and Applications in Mechanical Engineering and Mechatronics* (pp. 154–176). Hershey, PA: IGI Global. doi:10.4018/978-1-5225-1639-2.ch008

Henao, J., Poblano-Salas, C. A., Vargas, F., Giraldo-Betancur, A. L., Corona-Castuera, J., & Sotelo-Mazón, O. (2021). Principles and Applications of Thermal Spray Coatings. In S. Roy & G. Bose (Eds.), *Advanced Surface Coating Techniques for Modern Industrial Applications* (pp. 31–70). IGI Global. https://doi.org/10.4018/978-1-7998-4870-7.ch002

Henao, J., & Sotelo, O. (2018). Surface Engineering at High Temperature: Thermal Cycling and Corrosion Resistance. In A. Pakseresht (Ed.), *Production, Properties, and Applications of High Temperature Coatings* (pp. 131–159). Hershey, PA: IGI Global. doi:10.4018/978-1-5225-4194-3.ch006

Hrnčič, M. K., Cör, D., & Knez, Ž. (2021). Supercritical Fluids as a Tool for Green Energy and Chemicals. In L. Chen (Ed.), *Handbook of Research on Advancements in Supercritical Fluids Applications for Sustainable Energy Systems* (pp. 761–791). IGI Global. doi:10.4018/978-1-7998-5796-9.ch021

Ibrahim, O., Erdem, S., & Gurbuz, E. (2021). Studying Physical and Chemical Properties of Graphene Oxide and Reduced Graphene Oxide and Their Applications in Sustainable Building Materials. In R. González-Lezcano (Ed.), *Advancements in Sustainable Architecture and Energy Efficiency* (pp. 221–238). IGI Global. https://doi.org/10.4018/978-1-7998-7023-4.ch010

Ihianle, I. K., Islam, S., Naeem, U., & Ebenuwa, S. H. (2021). Exploiting Patterns of Object Use for Human Activity Recognition. In A. Nwajana & I. Ihianle (Eds.), *Handbook of Research on 5G Networks and Advancements in Computing, Electronics, and Electrical Engineering* (pp. 382–401). IGI Global. https://doi.org/10.4018/978-1-7998-6992-4.ch015

Ijemaru, G. K., Ngharamike, E. T., Oleka, E. U., & Nwajana, A. O. (2021). An Energy-Efficient Model for Opportunistic Data Collection in IoV-Enabled SC Waste Management. In A. Nwajana & I. Ihianle (Eds.), *Handbook of Research on 5G Networks and Advancements in Computing, Electronics, and Electrical Engineering* (pp. 1–19). IGI Global. https://doi.org/10.4018/978-1-7998-6992-4.ch001

Ilori, O. O., Adetan, D. A., & Umoru, L. E. (2017). Effect of Cutting Parameters on the Surface Residual Stress of Face-Milled Pearlitic Ductile Iron. *International Journal of Materials Forming and Machining Processes, 4*(1), 38–52. doi:10.4018/IJMFMP.2017010103

Imam, M. H., Tasadduq, I. A., Ahmad, A., Aldosari, F., & Khan, H. (2017). Automated Generation of Course Improvement Plans Using Expert System. *International Journal of Quality Assurance in Engineering and Technology Education*, *6*(1), 1–12. doi:10.4018/IJQAETE.2017010101

Injeti, S. K., & Kumar, T. V. (2018). A WDO Framework for Optimal Deployment of DGs and DSCs in a Radial Distribution System Under Daily Load Pattern to Improve Techno-Economic Benefits. *International Journal of Energy Optimization and Engineering*, *7*(2), 1–38. doi:10.4018/IJEOE.2018040101

Ishii, N., Anami, K., & Knisely, C. W. (2018). *Dynamic Stability of Hydraulic Gates and Engineering for Flood Prevention*. Hershey, PA: IGI Global. doi:10.4018/978-1-5225-3079-4

Iwamoto, Y., & Yamaguchi, H. (2021). Application of Supercritical Carbon Dioxide for Solar Water Heater. In L. Chen (Ed.), *Handbook of Research on Advancements in Supercritical Fluids Applications for Sustainable Energy Systems* (pp. 370–387). IGI Global. https://doi.org/10.4018/978-1-7998-5796-9.ch010

Jayapalan, S. (2018). A Review of Chemical Treatments on Natural Fibers-Based Hybrid Composites for Engineering Applications. In K. Kumar & J. Davim (Eds.), *Composites and Advanced Materials for Industrial Applications* (pp. 16–37). Hershey, PA: IGI Global. doi:10.4018/978-1-5225-5216-1.ch002

Kapetanakis, T. N., Vardiambasis, I. O., Ioannidou, M. P., & Konstantaras, A. I. (2021). Modeling Antenna Radiation Using Artificial Intelligence Techniques: The Case of a Circular Loop Antenna. In C. Nikolopoulos (Ed.), *Recent Trends on Electromagnetic Environmental Effects for Aeronautics and Space Applications* (pp. 186–225). IGI Global. https://doi.org/10.4018/978-1-7998-4879-0.ch007

Karkalos, N. E., Markopoulos, A. P., & Dossis, M. F. (2017). Optimal Model Parameters of Inverse Kinematics Solution of a 3R Robotic Manipulator Using ANN Models. *International Journal of Manufacturing, Materials, and Mechanical Engineering*, *7*(3), 20–40. doi:10.4018/IJMMME.2017070102

Kelly, M., Costello, M., Nicholson, G., & O'Connor, J. (2021). The Evolving Integration of BIM Into Built Environment Programmes in a Higher Education Institute. In J. Underwood & M. Shelbourn (Eds.), *Handbook of Research on Driving Transformational Change in the Digital Built Environment* (pp. 294–326). IGI Global. https://doi.org/10.4018/978-1-7998-6600-8.ch012

Kesimal, A., Karaman, K., Cihangir, F., & Ercikdi, B. (2018). Excavatability Assessment of Rock Masses for Geotechnical Studies. In N. Ceryan (Ed.), *Handbook of Research on Trends and Digital Advances in Engineering Geology* (pp. 231–256). Hershey, PA: IGI Global. doi:10.4018/978-1-5225-2709-1.ch006

Knoflacher, H. (2017). The Role of Engineers and Their Tools in the Transport Sector after Paradigm Change: From Assumptions and Extrapolations to Science. In H. Knoflacher & E. Ocalir-Akunal (Eds.), *Engineering Tools and Solutions for Sustainable Transportation Planning* (pp. 1–29). Hershey, PA: IGI Global. doi:10.4018/978-1-5225-2116-7.ch001

Kose, U. (2018). Towards an Intelligent Biomedical Engineering With Nature-Inspired Artificial Intelligence Techniques. In U. Kose, G. Guraksin, & O. Deperlioglu (Eds.), *Nature-Inspired Intelligent Techniques for Solving Biomedical Engineering Problems* (pp. 1–26). Hershey, PA: IGI Global. doi:10.4018/978-1-5225-4769-3.ch001

Kostić, S. (2018). A Review on Enhanced Stability Analyses of Soil Slopes Using Statistical Design. In N. Ceryan (Ed.), *Handbook of Research on Trends and Digital Advances in Engineering Geology* (pp. 446–481). Hershey, PA: IGI Global. doi:10.4018/978-1-5225-2709-1.ch013

Kumar, A., Patil, P. P., & Prajapati, Y. K. (2018). *Advanced Numerical Simulations in Mechanical Engineering*. Hershey, PA: IGI Global. doi:10.4018/978-1-5225-3722-9

Kumar, G. R., Rajyalakshmi, G., & Manupati, V. K. (2017). Surface Micro Patterning of Aluminium Reinforced Composite through Laser Peening. *International Journal of Manufacturing, Materials, and Mechanical Engineering*, *7*(4), 15–27. doi:10.4018/IJMMME.2017100102

Kumar, N., Basu, D. N., & Chen, L. (2021). Effect of Flow Acceleration and Buoyancy on Thermalhydraulics of sCO2 in Mini/Micro-Channel. In L. Chen (Ed.), *Handbook of Research on Advancements in Supercritical Fluids Applications for Sustainable Energy Systems* (pp. 161–182). IGI Global. doi:10.4018/978-1-7998-5796-9.ch005

Kumari, N., & Kumar, K. (2018). Fabrication of Orthotic Calipers With Epoxy-Based Green Composite. In K. Kumar & J. Davim (Eds.), *Composites and Advanced Materials for Industrial Applications* (pp. 157–176). Hershey, PA: IGI Global. doi:10.4018/978-1-5225-5216-1.ch008

Kuppusamy, R. R. (2018). Development of Aerospace Composite Structures Through Vacuum-Enhanced Resin Transfer Moulding Technology (VERTMTy): Vacuum-Enhanced Resin Transfer Moulding. In K. Kumar & J. Davim (Eds.), *Composites and Advanced Materials for Industrial Applications* (pp. 99–111). Hershey, PA: IGI Global. doi:10.4018/978-1-5225-5216-1.ch005

Kurganov, V. A., Zeigarnik, Y. A., & Maslakova, I. V. (2021). Normal and Deteriorated Heat Transfer Under Heating Turbulent Supercritical Pressure Coolants Flows in Round Tubes. In L. Chen (Ed.), *Handbook of Research on Advancements in Supercritical Fluids Applications for Sustainable Energy Systems* (pp. 494–532). IGI Global. https://doi.org/10.4018/978-1-7998-5796-9.ch014

Li, H., & Zhang, Y. (2021). Heat Transfer and Fluid Flow Modeling for Supercritical Fluids in Advanced Energy Systems. In L. Chen (Ed.), *Handbook of Research on Advancements in Supercritical Fluids Applications for Sustainable Energy Systems* (pp. 388–422). IGI Global. https://doi.org/10.4018/978-1-7998-5796-9.ch011

Loy, J., Howell, S., & Cooper, R. (2017). Engineering Teams: Supporting Diversity in Engineering Education. In M. Gray & K. Thomas (Eds.), *Strategies for Increasing Diversity in Engineering Majors and Careers* (pp. 106–129). Hershey, PA: IGI Global. doi:10.4018/978-1-5225-2212-6.ch006

Macher, G., Armengaud, E., Kreiner, C., Brenner, E., Schmittner, C., Ma, Z., ... Krammer, M. (2018). Integration of Security in the Development Lifecycle of Dependable Automotive CPS. In N. Druml, A. Genser, A. Krieg, M. Menghin, & A. Hoeller (Eds.), *Solutions for Cyber-Physical Systems Ubiquity* (pp. 383–423). Hershey, PA: IGI Global. doi:10.4018/978-1-5225-2845-6.ch015

Madhu, M. N., Singh, J. G., Mohan, V., & Ongsakul, W. (2021). Transmission Risk Optimization in Interconnected Systems: Risk-Adjusted Available Transfer Capability. In P. Vasant, G. Weber, & W. Punurai (Ed.), *Research Advancements in Smart Technology, Optimization, and Renewable Energy* (pp. 183-199). IGI Global. https://doi.org/10.4018/978-1-7998-3970-5.ch010

Mahendramani, G., & Lakshmana Swamy, N. (2018). Effect of Weld Groove Area on Distortion of Butt Welded Joints in Submerged Arc Welding. *International Journal of Manufacturing, Materials, and Mechanical Engineering*, *8*(2), 33–44. doi:10.4018/IJMMME.2018040103

Makropoulos, G., Koumaras, H., Setaki, F., Filis, K., Lutz, T., Montowtt, P., Tomaszewski, L., Dybiec, P., & Järvet, T. (2021). 5G and Unmanned Aerial Vehicles (UAVs) Use Cases: Analysis of the Ecosystem, Architecture, and Applications. In A. Nwajana & I. Ihianle (Eds.), *Handbook of Research on 5G Networks and Advancements in Computing, Electronics, and Electrical Engineering* (pp. 36–69). IGI Global. https://doi.org/10.4018/978-1-7998-6992-4.ch003

Meric, E. M., Erdem, S., & Gurbuz, E. (2021). Application of Phase Change Materials in Construction Materials for Thermal Energy Storage Systems in Buildings. In R. González-Lezcano (Ed.), *Advancements in Sustainable Architecture and Energy Efficiency* (pp. 1–20). IGI Global. https://doi.org/10.4018/978-1-7998-7023-4.ch001

Mihret, E. T., & Yitayih, K. A. (2021). Operation of VANET Communications: The Convergence of UAV System With LTE/4G and WAVE Technologies. *International Journal of Smart Vehicles and Smart Transportation*, *4*(1), 29–51. https://doi.org/10.4018/IJSVST.2021010103

Mir, M. A., Bhat, B. A., Sheikh, B. A., Rather, G. A., Mehraj, S., & Mir, W. R. (2021). Nanomedicine in Human Health Therapeutics and Drug Delivery: Nanobiotechnology and Nanobiomedicine. In M. Bhat, I. Wani, & S. Ashraf (Eds.), *Applications of Nanomaterials in Agriculture, Food Science, and Medicine* (pp. 229–251). IGI Global. doi:10.4018/978-1-7998-5563-7.ch013

Mohammadzadeh, S., & Kim, Y. (2017). Nonlinear System Identification of Smart Buildings. In P. Samui, S. Chakraborty, & D. Kim (Eds.), *Modeling and Simulation Techniques in Structural Engineering* (pp. 328–347). Hershey, PA: IGI Global. doi:10.4018/978-1-5225-0588-4.ch011

Molina, G. J., Aktaruzzaman, F., Soloiu, V., & Rahman, M. (2017). Design and Testing of a Jet-Impingement Instrument to Study Surface-Modification Effects by Nanofluids. *International Journal of Surface Engineering and Interdisciplinary Materials Science*, *5*(2), 43–61. doi:10.4018/IJSEIMS.2017070104

Moreno-Rangel, A., & Carrillo, G. (2021). Energy-Efficient Homes: A Heaven for Respiratory Illnesses. In R. González-Lezcano (Ed.), *Advancements in Sustainable Architecture and Energy Efficiency* (pp. 49–71). IGI Global. https://doi.org/10.4018/978-1-7998-7023-4.ch003

Msomi, V., & Jantjies, B. T. (2021). Correlative Analysis Between Tensile Properties and Tool Rotational Speeds of Friction Stir Welded Similar Aluminium Alloy Joints. *International Journal of Surface Engineering and Interdisciplinary Materials Science*, *9*(2), 58–78. https://doi.org/10.4018/IJSEIMS.2021070104

Muigai, M. N., Mwema, F. M., Akinlabi, E. T., & Obiko, J. O. (2021). Surface Engineering of Materials Through Weld-Based Technologies: An Overview. In S. Roy & G. Bose (Eds.), *Advanced Surface Coating Techniques for Modern Industrial Applications* (pp. 247–260). IGI Global. doi:10.4018/978-1-7998-4870-7.ch011

Mukherjee, A., Saeed, R. A., Dutta, S., & Naskar, M. K. (2017). Fault Tracking Framework for Software-Defined Networking (SDN). In C. Singhal & S. De (Eds.), *Resource Allocation in Next-Generation Broadband Wireless Access Networks* (pp. 247–272). Hershey, PA: IGI Global. doi:10.4018/978-1-5225-2023-8.ch011

Mukhopadhyay, A., Barman, T. K., & Sahoo, P. (2018). Electroless Nickel Coatings for High Temperature Applications. In K. Kumar & J. Davim (Eds.), *Composites and Advanced Materials for Industrial Applications* (pp. 297–331). Hershey, PA: IGI Global. doi:10.4018/978-1-5225-5216-1.ch013

Mwema, F. M., & Wambua, J. M. (2022). Machining of Poly Methyl Methacrylate (PMMA) and Other Olymeric Materials: A Review. In K. Kumar, B. Babu, & J. Davim (Eds.), *Handbook of Research on Advancements in the Processing, Characterization, and Application of Lightweight Materials* (pp. 363–379). IGI Global. https://doi.org/10.4018/978-1-7998-7864-3.ch016

Mykhailyshyn, R., Savkiv, V., Boyko, I., Prada, E., & Virgala, I. (2021). Substantiation of Parameters of Friction Elements of Bernoulli Grippers With a Cylindrical Nozzle. *International Journal of Manufacturing, Materials, and Mechanical Engineering*, *11*(2), 17–39. https://doi.org/10.4018/IJMMME.2021040102

Náprstek, J., & Fischer, C. (2017). Dynamic Stability and Post-Critical Processes of Slender Auto-Parametric Systems. In V. Plevris, G. Kremmyda, & Y. Fahjan (Eds.), *Performance-Based Seismic Design of Concrete Structures and Infrastructures* (pp. 128–171). Hershey, PA: IGI Global. doi:10.4018/978-1-5225-2089-4.ch006

Nautiyal, L., Shivach, P., & Ram, M. (2018). Optimal Designs by Means of Genetic Algorithms. In M. Ram & J. Davim (Eds.), *Soft Computing Techniques and Applications in Mechanical Engineering* (pp. 151–161). Hershey, PA: IGI Global. doi:10.4018/978-1-5225-3035-0.ch007

Nazir, R. (2017). Advanced Nanomaterials for Water Engineering and Treatment: Nano-Metal Oxides and Their Nanocomposites. In T. Saleh (Ed.), *Advanced Nanomaterials for Water Engineering, Treatment, and Hydraulics* (pp. 84–126). Hershey, PA: IGI Global. doi:10.4018/978-1-5225-2136-5.ch005

Nikolopoulos, C. D. (2021). Recent Advances on Measuring and Modeling ELF-Radiated Emissions for Space Applications. In C. Nikolopoulos (Ed.), *Recent Trends on Electromagnetic Environmental Effects for Aeronautics and Space Applications* (pp. 1–38). IGI Global. https://doi.org/10.4018/978-1-7998-4879-0.ch001

Nogueira, A. F., Ribeiro, J. C., Fernández de Vega, F., & Zenha-Rela, M. A. (2018). Evolutionary Approaches to Test Data Generation for Object-Oriented Software: Overview of Techniques and Tools. In M. Khosrow-Pour, D.B.A. (Ed.), Incorporating Nature-Inspired Paradigms in Computational Applications (pp. 162-194). Hershey, PA: IGI Global. https://doi.org/ doi:10.4018/978-1-5225-5020-4.ch006

Nwajana, A. O., Obi, E. R., Ijemaru, G. K., Oleka, E. U., & Anthony, D. C. (2021). Fundamentals of RF/Microwave Bandpass Filter Design. In A. Nwajana & I. Ihianle (Eds.), *Handbook of Research on 5G Networks and Advancements in Computing, Electronics, and Electrical Engineering* (pp. 149–164). IGI Global. https://doi.org/10.4018/978-1-7998-6992-4.ch005

Ogbodo, E. A. (2021). Comparative Study of Transmission Line Junction vs. Asynchronously Coupled Junction Diplexers. In A. Nwajana & I. Ihianle (Eds.), *Handbook of Research on 5G Networks and Advancements in Computing, Electronics, and Electrical Engineering* (pp. 326–336). IGI Global. https://doi.org/10.4018/978-1-7998-6992-4.ch013

Orosa, J. A., Vergara, D., Fraguela, F., & Masdías-Bonome, A. (2021). Statistical Understanding and Optimization of Building Energy Consumption and Climate Change Consequences. In R. González-Lezcano (Ed.), *Advancements in Sustainable Architecture and Energy Efficiency* (pp. 195–220). IGI Global. https://doi.org/10.4018/978-1-7998-7023-4.ch009

Osho, M. B. (2018). Industrial Enzyme Technology: Potential Applications. In S. Bharati & P. Chaurasia (Eds.), *Research Advancements in Pharmaceutical, Nutritional, and Industrial Enzymology* (pp. 375–394). Hershey, PA: IGI Global. doi:10.4018/978-1-5225-5237-6.ch017

Ouadi, A., & Zitouni, A. (2021). Phasor Measurement Improvement Using Digital Filter in a Smart Grid. In A. Recioui & H. Bentarzi (Eds.), *Optimizing and Measuring Smart Grid Operation and Control* (pp. 100–117). IGI Global. https://doi.org/10.4018/978-1-7998-4027-5.ch005

Padmaja, P., & Marutheswar, G. (2017). Certain Investigation on Secured Data Transmission in Wireless Sensor Networks. *International Journal of Mobile Computing and Multimedia Communications*, *8*(1), 48–61. doi:10.4018/IJMCMC.2017010104

Palmer, S., & Hall, W. (2017). An Evaluation of Group Work in First-Year Engineering Design Education. In R. Tucker (Ed.), *Collaboration and Student Engagement in Design Education* (pp. 145–168). Hershey, PA: IGI Global. doi:10.4018/978-1-5225-0726-0.ch007

Panchenko, V. (2021). Prospects for Energy Supply of the Arctic Zone Objects of Russia Using Frost-Resistant Solar Modules. In P. Vasant, G. Weber, & W. Punurai (Eds.), *Research Advancements in Smart Technology, Optimization, and Renewable Energy* (pp. 149-169). IGI Global. https://doi.org/10.4018/978-1-7998-3970-5.ch008

Panchenko, V. (2021). Photovoltaic Thermal Module With Paraboloid Type Solar Concentrators. *International Journal of Energy Optimization and Engineering*, *10*(2), 1–23. https://doi.org/10.4018/IJEOE.2021040101

Pandey, K., & Datta, S. (2021). Dry Machining of Inconel 825 Superalloys: Performance of Tool Inserts (Carbide, Cermet, and SiAlON). *International Journal of Manufacturing, Materials, and Mechanical Engineering, 11*(4), 26–39. doi:10.4018/IJMMME.2021100102

Panneer, R. (2017). Effect of Composition of Fibers on Properties of Hybrid Composites. *International Journal of Manufacturing, Materials, and Mechanical Engineering, 7*(4), 28–43. doi:10.4018/IJMMME.2017100103

Pany, C. (2021). Estimation of Correct Long-Seam Mismatch Using FEA to Compare the Measured Strain in a Non-Destructive Testing of a Pressurant Tank: A Reverse Problem. *International Journal of Smart Vehicles and Smart Transportation, 4*(1), 16–28. doi:10.4018/IJSVST.2021010102

Paul, S., & Roy, P. (2018). Optimal Design of Power System Stabilizer Using a Novel Evolutionary Algorithm. *International Journal of Energy Optimization and Engineering, 7*(3), 24–46. doi:10.4018/IJEOE.2018070102

Paul, S., & Roy, P. K. (2021). Oppositional Differential Search Algorithm for the Optimal Tuning of Both Single Input and Dual Input Power System Stabilizer. In P. Vasant, G. Weber, & W. Punurai (Eds.), *Research Advancements in Smart Technology, Optimization, and Renewable Energy* (pp. 256-282). IGI Global. https://doi.org/10.4018/978-1-7998-3970-5.ch013

Pavaloiu, A. (2018). Artificial Intelligence Ethics in Biomedical-Engineering-Oriented Problems. In U. Kose, G. Guraksin, & O. Deperlioglu (Eds.), *Nature-Inspired Intelligent Techniques for Solving Biomedical Engineering Problems* (pp. 219–231). Hershey, PA: IGI Global. doi:10.4018/978-1-5225-4769-3.ch010

Pioro, I., Mahdi, M., & Popov, R. (2017). Application of Supercritical Pressures in Power Engineering. In L. Chen & Y. Iwamoto (Eds.), *Advanced Applications of Supercritical Fluids in Energy Systems* (pp. 404–457). Hershey, PA: IGI Global. doi:10.4018/978-1-5225-2047-4.ch013

Plaksina, T., & Gildin, E. (2017). Rigorous Integrated Evolutionary Workflow for Optimal Exploitation of Unconventional Gas Assets. *International Journal of Energy Optimization and Engineering, 6*(1), 101–122. doi:10.4018/IJEOE.2017010106

Popat, J., Kakadiya, H., Tak, L., Singh, N. K., Majeed, M. A., & Mahajan, V. (2021). Reliability of Smart Grid Including Cyber Impact: A Case Study. In R. Singh, A. Singh, A. Dwivedi, & P. Nagabhushan (Eds.), *Computational Methodologies for Electrical and Electronics Engineers* (pp. 163–174). IGI Global. https://doi.org/10.4018/978-1-7998-3327-7.ch013

Quiza, R., La Fé-Perdomo, I., Rivas, M., & Ramtahalsing, V. (2021). Triple Bottom Line-Focused Optimization of Oblique Turning Processes Based on Hybrid Modeling: A Study Case on AISI 1045 Steel Turning. In L. Burstein (Ed.), *Handbook of Research on Advancements in Manufacturing, Materials, and Mechanical Engineering* (pp. 215–241). IGI Global. https://doi.org/10.4018/978-1-7998-4939-1.ch010

Rahmani, M. K. (2022). Blockchain Technology: Principles and Algorithms. In S. Khan, M. Syed, R. Hammad, & A. Bushager (Eds.), *Blockchain Technology and Computational Excellence for Society 5.0* (pp. 16–27). IGI Global. https://doi.org/10.4018/978-1-7998-8382-1.ch002

Ramdani, N., & Azibi, M. (2018). Polymer Composite Materials for Microelectronics Packaging Applications: Composites for Microelectronics Packaging. In K. Kumar & J. Davim (Eds.), *Composites and Advanced Materials for Industrial Applications* (pp. 177–211). Hershey, PA: IGI Global. doi:10.4018/978-1-5225-5216-1.ch009

Ramesh, M., Garg, R., & Subrahmanyam, G. V. (2017). Investigation of Influence of Quenching and Annealing on the Plane Fracture Toughness and Brittle to Ductile Transition Temperature of the Zinc Coated Structural Steel Materials. *International Journal of Surface Engineering and Interdisciplinary Materials Science*, *5*(2), 33–42. doi:10.4018/IJSEIMS.2017070103

Robinson, J., & Beneroso, D. (2022). Project-Based Learning in Chemical Engineering: Curriculum and Assessment, Culture and Learning Spaces. In A. Alves & N. van Hattum-Janssen (Eds.), *Training Engineering Students for Modern Technological Advancement* (pp. 1–19). IGI Global. https://doi.org/10.4018/978-1-7998-8816-1.ch001

Rondon, B. (2021). Experimental Characterization of Admittance Meter With Crude Oil Emulsions. *International Journal of Electronics, Communications, and Measurement Engineering*, *10*(2), 51–59. https://doi.org/10.4018/IJECME.2021070104

Rudolf, S., Biryuk, V. V., & Volov, V. (2018). Vortex Effect, Vortex Power: Technology of Vortex Power Engineering. In V. Kharchenko & P. Vasant (Eds.), *Handbook of Research on Renewable Energy and Electric Resources for Sustainable Rural Development* (pp. 500–533). Hershey, PA: IGI Global. doi:10.4018/978-1-5225-3867-7.ch021

Sah, A., Bhadula, S. J., Dumka, A., & Rawat, S. (2018). A Software Engineering Perspective for Development of Enterprise Applications. In A. Elçi (Ed.), *Handbook of Research on Contemporary Perspectives on Web-Based Systems* (pp. 1–23). Hershey, PA: IGI Global. doi:10.4018/978-1-5225-5384-7.ch001

Sahli, Y., Zitouni, B., & Hocine, B. M. (2021). Three-Dimensional Numerical Study of Overheating of Two Intermediate Temperature P-AS-SOFC Geometrical Configurations. In G. Badea, R. Felseghi, & I. Aşchilean (Eds.), *Hydrogen Fuel Cell Technology for Stationary Applications* (pp. 186–222). IGI Global. https://doi.org/10.4018/978-1-7998-4945-2.ch008

Sahoo, P., & Roy, S. (2017). Tribological Behavior of Electroless Ni-P, Ni-P-W and Ni-P-Cu Coatings: A Comparison. *International Journal of Surface Engineering and Interdisciplinary Materials Science*, *5*(1), 1–15. doi:10.4018/IJSEIMS.2017010101

Sahoo, S. (2018). Laminated Composite Hypar Shells as Roofing Units: Static and Dynamic Behavior. In K. Kumar & J. Davim (Eds.), *Composites and Advanced Materials for Industrial Applications* (pp. 249–269). Hershey, PA: IGI Global. doi:10.4018/978-1-5225-5216-1.ch011

Sahu, H., & Hungyo, M. (2018). Introduction to SDN and NFV. In A. Dumka (Ed.), *Innovations in Software-Defined Networking and Network Functions Virtualization* (pp. 1–25). Hershey, PA: IGI Global. doi:10.4018/978-1-5225-3640-6.ch001

Salem, A. M., & Shmelova, T. (2018). Intelligent Expert Decision Support Systems: Methodologies, Applications, and Challenges. In T. Shmelova, Y. Sikirda, N. Rizun, A. Salem, & Y. Kovalyov (Eds.), *Socio-Technical Decision Support in Air Navigation Systems: Emerging Research and Opportunities* (pp. 215–242). Hershey, PA: IGI Global. doi:10.4018/978-1-5225-3108-1.ch007

Samal, M. (2017). FE Analysis and Experimental Investigation of Cracked and Un-Cracked Thin-Walled Tubular Components to Evaluate Mechanical and Fracture Properties. In P. Samui, S. Chakraborty, & D. Kim (Eds.), *Modeling and Simulation Techniques in Structural Engineering* (pp. 266–293). Hershey, PA: IGI Global. doi:10.4018/978-1-5225-0588-4.ch009

Samal, M., & Balakrishnan, K. (2017). Experiments on a Ring Tension Setup and FE Analysis to Evaluate Transverse Mechanical Properties of Tubular Components. In P. Samui, S. Chakraborty, & D. Kim (Eds.), *Modeling and Simulation Techniques in Structural Engineering* (pp. 91–115). Hershey, PA: IGI Global. doi:10.4018/978-1-5225-0588-4.ch004

Samarasinghe, D. A., & Wood, E. (2021). Innovative Digital Technologies. In J. Underwood & M. Shelbourn (Eds.), *Handbook of Research on Driving Transformational Change in the Digital Built Environment* (pp. 142–163). IGI Global. https://doi.org/10.4018/978-1-7998-6600-8.ch006

Sawant, S. (2018). Deep Learning and Biomedical Engineering. In U. Kose, G. Guraksin, & O. Deperlioglu (Eds.), *Nature-Inspired Intelligent Techniques for Solving Biomedical Engineering Problems* (pp. 283–296). Hershey, PA: IGI Global. doi:10.4018/978-1-5225-4769-3.ch014

Schulenberg, T. (2021). Energy Conversion Using the Supercritical Steam Cycle. In L. Chen (Ed.), *Handbook of Research on Advancements in Supercritical Fluids Applications for Sustainable Energy Systems* (pp. 659–681). IGI Global. doi:10.4018/978-1-7998-5796-9.ch018

Sezgin, H., & Berkalp, O. B. (2018). Textile-Reinforced Composites for the Automotive Industry. In K. Kumar & J. Davim (Eds.), *Composites and Advanced Materials for Industrial Applications* (pp. 129–156). Hershey, PA: IGI Global. doi:10.4018/978-1-5225-5216-1.ch007

Shaaban, A. A., & Shehata, O. M. (2021). Combining Response Surface Method and Metaheuristic Algorithms for Optimizing SPIF Process. *International Journal of Manufacturing, Materials, and Mechanical Engineering, 11*(4), 1–25. https://doi.org/10.4018/IJMMME.2021100101

Shafaati Shemami, M., & Sefid, M. (2022). Implementation and Demonstration of Electric Vehicle-to-Home (V2H) Application: A Case Study. In M. Alam, R. Pillai, & N. Murugesan (Eds.), *Developing Charging Infrastructure and Technologies for Electric Vehicles* (pp. 268–293). IGI Global. https://doi.org/10.4018/978-1-7998-6858-3.ch015

Shah, M. Z., Gazder, U., Bhatti, M. S., & Hussain, M. (2018). Comparative Performance Evaluation of Effects of Modifier in Asphaltic Concrete Mix. *International Journal of Strategic Engineering*, *1*(2), 13–25. doi:10.4018/IJoSE.2018070102

Sharma, N., & Kumar, K. (2018). Fabrication of Porous NiTi Alloy Using Organic Binders. In K. Kumar & J. Davim (Eds.), *Composites and Advanced Materials for Industrial Applications* (pp. 38–62). Hershey, PA: IGI Global. doi:10.4018/978-1-5225-5216-1.ch003

Shivach, P., Nautiyal, L., & Ram, M. (2018). Applying Multi-Objective Optimization Algorithms to Mechanical Engineering. In M. Ram & J. Davim (Eds.), *Soft Computing Techniques and Applications in Mechanical Engineering* (pp. 287–301). Hershey, PA: IGI Global. doi:10.4018/978-1-5225-3035-0.ch014

Shmelova, T. (2018). Stochastic Methods for Estimation and Problem Solving in Engineering: Stochastic Methods of Decision Making in Aviation. In S. Kadry (Ed.), *Stochastic Methods for Estimation and Problem Solving in Engineering* (pp. 139–160). Hershey, PA: IGI Global. doi:10.4018/978-1-5225-5045-7.ch006

Siero González, L. R., & Romo Vázquez, A. (2017). Didactic Sequences Teaching Mathematics for Engineers With Focus on Differential Equations. In M. Ramírez-Montoya (Ed.), *Handbook of Research on Driving STEM Learning With Educational Technologies* (pp. 129–151). Hershey, PA: IGI Global. doi:10.4018/978-1-5225-2026-9.ch007

Sim, M. S., You, K. Y., Esa, F., & Chan, Y. L. (2021). Nanostructured Electromagnetic Metamaterials for Sensing Applications. In M. Bhat, I. Wani, & S. Ashraf (Eds.), *Applications of Nanomaterials in Agriculture, Food Science, and Medicine* (pp. 141–164). IGI Global. https://doi.org/10.4018/978-1-7998-5563-7.ch009

Singh, R., & Dutta, S. (2018). Visible Light Active Nanocomposites for Photocatalytic Applications. In K. Kumar & J. Davim (Eds.), *Composites and Advanced Materials for Industrial Applications* (pp. 270–296). Hershey, PA: IGI Global. doi:10.4018/978-1-5225-5216-1.ch012

Skripov, P. V., Yampol'skiy, A. D., & Rutin, S. B. (2021). High-Power Heat Transfer in Supercritical Fluids: Microscale Times and Sizes. In L. Chen (Ed.), *Handbook of Research on Advancements in Supercritical Fluids Applications for Sustainable Energy Systems* (pp. 424–450). IGI Global. https://doi.org/10.4018/978-1-7998-5796-9.ch012

Sözbilir, H., Özkaymak, Ç., Uzel, B., & Sümer, Ö. (2018). Criteria for Surface Rupture Microzonation of Active Faults for Earthquake Hazards in Urban Areas. In N. Ceryan (Ed.), *Handbook of Research on Trends and Digital Advances in Engineering Geology* (pp. 187–230). Hershey, PA: IGI Global. doi:10.4018/978-1-5225-2709-1.ch005

Stanciu, I. (2018). Stochastic Methods in Microsystems Engineering. In S. Kadry (Ed.), *Stochastic Methods for Estimation and Problem Solving in Engineering* (pp. 161–176). Hershey, PA: IGI Global. doi:10.4018/978-1-5225-5045-7.ch007

Strebkov, D., Nekrasov, A., Trubnikov, V., & Nekrasov, A. (2018). Single-Wire Resonant Electric Power Systems for Renewable-Based Electric Grid. In V. Kharchenko & P. Vasant (Eds.), *Handbook of Research on Renewable Energy and Electric Resources for Sustainable Rural Development* (pp. 449–474). Hershey, PA: IGI Global. doi:10.4018/978-1-5225-3867-7.ch019

Sukhyy, K., Belyanovskaya, E., & Sukhyy, M. (2021). *Basic Principles for Substantiation of Working Pair Choice*. IGI Global. doi:10.4018/978-1-7998-4432-7.ch002

Suri, M. S., & Kaliyaperumal, D. (2022). Extension of Aspiration Level Model for Optimal Planning of Fast Charging Stations. In A. Fekik & N. Benamrouche (Eds.), *Modeling and Control of Static Converters for Hybrid Storage Systems* (pp. 91–106). IGI Global. https://doi.org/10.4018/978-1-7998-7447-8.ch004

Tallet, E., Gledson, B., Rogage, K., Thompson, A., & Wiggett, D. (2021). Digitally-Enabled Design Management. In J. Underwood & M. Shelbourn (Eds.), *Handbook of Research on Driving Transformational Change in the Digital Built Environment* (pp. 63–89). IGI Global. https://doi.org/10.4018/978-1-7998-6600-8.ch003

Terki, A., & Boubertakh, H. (2021). A New Hybrid Binary-Real Coded Cuckoo Search and Tabu Search Algorithm for Solving the Unit-Commitment Problem. *International Journal of Energy Optimization and Engineering, 10*(2), 104–119. https://doi.org/10.4018/IJEOE.2021040105

Tüdeş, Ş., Kumlu, K. B., & Ceryan, S. (2018). Integration Between Urban Planning and Natural Hazards For Resilient City. In N. Ceryan (Ed.), *Handbook of Research on Trends and Digital Advances in Engineering Geology* (pp. 591–630). Hershey, PA: IGI Global. doi:10.4018/978-1-5225-2709-1.ch017

Ulamis, K. (2018). Soil Liquefaction Assessment by Anisotropic Cyclic Triaxial Test. In N. Ceryan (Ed.), *Handbook of Research on Trends and Digital Advances in Engineering Geology* (pp. 631–664). Hershey, PA: IGI Global. doi:10.4018/978-1-5225-2709-1.ch018

Valente, M., & Milani, G. (2017). Seismic Assessment and Retrofitting of an Under-Designed RC Frame Through a Displacement-Based Approach. In V. Plevris, G. Kremmyda, & Y. Fahjan (Eds.), *Performance-Based Seismic Design of Concrete Structures and Infrastructures* (pp. 36–58). Hershey, PA: IGI Global. doi:10.4018/978-1-5225-2089-4.ch002

Vargas-Bernal, R. (2021). Advances in Electromagnetic Environmental Shielding for Aeronautics and Space Applications. In C. Nikolopoulos (Ed.), *Recent Trends on Electromagnetic Environmental Effects for Aeronautics and Space Applications* (pp. 80–96). IGI Global. https://doi.org/10.4018/978-1-7998-4879-0.ch003

Vasant, P. (2018). A General Medical Diagnosis System Formed by Artificial Neural Networks and Swarm Intelligence Techniques. In U. Kose, G. Guraksin, & O. Deperlioglu (Eds.), *Nature-Inspired Intelligent Techniques for Solving Biomedical Engineering Problems* (pp. 130–145). Hershey, PA: IGI Global. doi:10.4018/978-1-5225-4769-3.ch006

Verner, C. M., & Sarwar, D. (2021). Avoiding Project Failure and Achieving Project Success in NHS IT System Projects in the United Kingdom. *International Journal of Strategic Engineering, 4*(1), 33–54. https://doi.org/10.4018/IJoSE.2021010103

Verrollot, J., Tolonen, A., Harkonen, J., & Haapasalo, H. J. (2018). Challenges and Enablers for Rapid Product Development. *International Journal of Applied Industrial Engineering, 5*(1), 25–49. doi:10.4018/IJAIE.2018010102

Wan, A. C., Zulu, S. L., & Khosrow-Shahi, F. (2021). Industry Views on BIM for Site Safety in Hong Kong. In J. Underwood & M. Shelbourn (Eds.), *Handbook of Research on Driving Transformational Change in the Digital Built Environment* (pp. 120–140). IGI Global. https://doi.org/10.4018/978-1-7998-6600-8.ch005

Yardımci, A. G., & Karpuz, C. (2018). Fuzzy Rock Mass Rating: Soft-Computing-Aided Preliminary Stability Analysis of Weak Rock Slopes. In N. Ceryan (Ed.), *Handbook of Research on Trends and Digital Advances in Engineering Geology* (pp. 97–131). Hershey, PA: IGI Global. doi:10.4018/978-1-5225-2709-1.ch003

You, K. Y. (2021). Development Electronic Design Automation for RF/Microwave Antenna Using MATLAB GUI. In A. Nwajana & I. Ihianle (Eds.), *Handbook of Research on 5G Networks and Advancements in Computing, Electronics, and Electrical Engineering* (pp. 70–148). IGI Global. https://doi.org/10.4018/978-1-7998-6992-4.ch004

Yousefi, Y., Gratton, P., & Sarwar, D. (2021). Investigating the Opportunities to Improve the Thermal Performance of a Case Study Building in London. *International Journal of Strategic Engineering*, *4*(1), 1–18. https://doi.org/10.4018/IJoSE.2021010101

Zindani, D., & Kumar, K. (2018). Industrial Applications of Polymer Composite Materials. In K. Kumar & J. Davim (Eds.), *Composites and Advanced Materials for Industrial Applications* (pp. 1–15). Hershey, PA: IGI Global. doi:10.4018/978-1-5225-5216-1.ch001

Zindani, D., Maity, S. R., & Bhowmik, S. (2018). A Decision-Making Approach for Material Selection of Polymeric Composite Bumper Beam. In K. Kumar & J. Davim (Eds.), *Composites and Advanced Materials for Industrial Applications* (pp. 112–128). Hershey, PA: IGI Global. doi:10.4018/978-1-5225-5216-1.ch006

About the Author

Rupam Bhaduri is an alumnus of NIT–Durgapur (An Institute of National Importance of GoI). His M.Tech. is in Control Systems from MIT–MAHE (An Institute of Eminence by GoI). During his Ph.D. he worked on various categories of Electromagnetic Levitation Systems on its Control Systems, Power Converters and Optimisation Techniques from NIT–Dgp with external mentor from IIT–Kharagpur. Earlier he worked in the Thermal Power Plant industry for several years and acquired exhaustive real time core industrial experience, where he worked exhaustively for the projects of BHEL-ISG, L&T-ECC, Bajaj Electricals and Schneider Electric. He is recipient of several Best Paper Awards and the very prestigious TATA RAO Award in the 28th INDIAN ENGINEERING CONGRESS. He is a member of confidential reviwer committee of few prestigious R&D agencies. Earlier he has worked as Co-PI for DST-SERB and IE(I) funded projects. Presently he is working for a very prestigious Indo-Canadian Collaborative Consortium of Projects (funded by Department of Science & Technology, Ministry of S&T, Govt. of India) by partnering with few IITs, NITs & CSIR Labs. He is reviewer for IET(UK), T&F(UK), Springer-Nature Journals.

Index

G

H

I

L

M

N

O

P

S

T

W

Z

www.ingramcontent.com/pod-product-compliance
Lightning Source LLC
LaVergne TN
LVHW081323300926
849161LV00008B/1858